GENETIC ENGINEERING OF MESENCHYMAL STEM CELLS

GENETIC ENGINEERING OF MESENCHYMAL STEM CELLS

Edited by

JAN A. NOLTA

Washington University School of Medicine,
St. Louis, MO, USA

 Springer

A C.I.P. Catalogue record for this book is available from the Library of Congress.

ISBN-10 1-4020-3935-2 (HB)
ISBN-13 978-1-4020-3935-5 (HB)
ISBN-10 1-4020-3959-X (e-book)
ISBN-13 978-1-4020-3959-1 (e-book)

Published by Springer,
P.O. Box 17, 3300 AA Dordrecht, The Netherlands.

www.springer.com

Cover art:
A mixed MSC/HSC culture from human umbilical cord blood. Courtesy of Todd Meyerrose,
Washington University School of Medicine

Printed on acid-free paper

Printed in the Netherlands.

CONTENTS

CHAPTER 1

MESENCHYMAL STEM CELL ENGINEERING AND TRANSPLANTATION

Introduction

F. AERTS AND G. WAGEMAKER

Department of Hematology, Erasmus Medical Centre, Rotterdam,
The Netherlands

1. BONE MARROW STROMA AND STROMAL CELLS

Bone marrow (BM) is considered to be one of the large and complex organs of the human body and is the most important site of hematopoiesis after birth [1]. It is composed of stroma, hematopoietic cords and sinusoidal capillaries. Under normal conditions, the production of blood is exactly adjusted to the organism's needs (homeostasis) and is both regulated and maintained by interaction with the bone marrow stromal compartment. This intriguing 3-dimensional meshwork is composed of reticular cells with phagocytic properties and formed by a loose network of reticular fibers. The extravascular spaces between the marrow sinuses contain the hematopoietic stem cells and their descendents. The mesenchymal elements that constitute the stroma of the bone marrow form a morphologically heterogeneous population and include a variety of cell types, such as macrophages, fibroblasts, endothelial cells, adipocytes, blanket cells, adventitial reticular cells, osteogenic and barrier cells [2].

A continuum of interacting cells is formed within the bone marrow space, which extends from the abluminal sides of blood vessels to the bone surfaces through the stromal cells scattered between the hematopoietic cells. This results in the physical and biological continuity of bone and marrow, cooperating to form a single organ [3], in which the stroma provides a suitable chemical environment for the developing hematopoietic cells. Stromal cells in the primitive non-hematopoietic marrow, which impress much like preosteoblasts are actively dividing. In contrast, the stromal cells of actively hematopoietic marrow are quiescent, although they continue to express the osteoblastic marker alkaline phosphatase at high levels [4].

J.A. Nolta (ed.), Genetic Engineering of Mesenchymal Stem Cells, 1–44.
© 2006 Springer. Printed in the Netherlands.

Formation of the marrow cavity and marrow stroma requires core-binding factor A1 (Cbfa1), a transcription factor critical for osteoblast differentiation and essential during development and post-natal bone formation [5]. Defects in Cbfa1 result in a complete lack of ossification and are therefore not compatible with life. In postnatal organisms, expression of cbfa1 is a basic property in clones and non-transformed lines of both human and murine marrow stromal cells [6]. However, expression of Cbfa1 alone does not necessarily imply an actual osteogenic capacity, when retransplanted *in vivo*, because osteogenic differentiation is also determined and influenced by environmental cues. Furthermore, expression of Cbfa1 transcripts does not prevent the ability to differentiate towards other, non-osteoblastic, phenotypes such as adipocytes or chondrocytes [6].

The term "marrow stromal cells" has been used previously to describe the not yet fully defined population of cells, which are plastic-adherent and form the supportive cell layer in *in vitro* long-term bone marrow cultures (LT-BMC). This term is generally restricted to non-hematopoietic cells of mesenchymal origin and does not include other contaminating cells, such as macrophages, endothelial cells, smooth muscle cells and adipocytes, which are also components of the adherent layer [7]. Some of the cell types, which grow in LT-BMC may however not be actual functional components of the stroma for *in vivo* hematopoiesis. Although these cells, that comprise a major component of the marrow stroma, are often considered to be "fibroblastic" in nature, they are distinct both in function and by phenotype, from other fibroblasts. In physiologic conditions, the hematopoietic microenvironment in healthy bone marrow contains sparsely collagen. The deposition of discrete amounts of collagen fibers (fibrosis) in the bone marrow is usually a response to injury and true fibroblasts (involved in deposition of collagen), which appear in certain pathological states, are therefore not considered to be standard components of the bone marrow stroma [8].

2. WHAT ARE MESENCHYMAL STEM CELLS?

Marrow stromal cells can be grown in culture, but it has been difficult to point out the cell types identified *in situ*, and correlate them with equivalent cells that grow in long-term cultures. Neither morphological nor serologic criteria have thus far proved sufficiently reliable in predicting their function [9, 10]. Some of the bone marrow-derived stromal cells in culture have the capability, under specific conditions, to differentiate into several distinct mesenchymal lineages [11]. Although self-renewing progenitors have been demonstrated for each of these lineages separately, increasing evidence has been presented, suggesting that these cells may actually be derived from a single common precursor.

Friedenstein et al. [12] described in 1970 the presence of fibroblast colony-forming cells (FCFC) in LT-BMC. Since then, these mesenchymal multipotential progenitors have been referred to by a number of different names. The commonly used terms colony forming unit-fibroblast (CFU-F) [13] and marrow stromal fibroblast (MSF) [14, 15] have now been gradually replaced and other, unfortunately, equally indistinct terms, like marrow stromal cells (MSC) [16] or mesenchymal progenitor cells (MPC) [17] have been introduced. The intermixture of these terms has been predominantly a result of the absence of specialized assays to accurately assess the functional properties of these progenitor cells and the lack of adequate markers, suitable for positive selection of these cells without the contamination or interference of other adherent cells.

In postnatal organisms, each tissue and organ is now generally perceived to contain a small sub-population of quiescent cells, which are, when stimulated by local cues, capable of self-renewal and of indefinite or at least prolonged proliferative potential. These cells also share the ability to give rise to a large group of daughter cells, each with a different direction of specialization [18]. Between the stem cell and its terminally differentiated progeny, intermediate populations of committed progenitors, the transit amplifying cells, are present and display a more limited proliferative capacity with a restricted differentiation potential [19]. Multipotential stem cells are capable of regenerating tissue after injury [20] and have the ability to leave their "tissue" niche and circulate in the peripheral blood [19, 21, 22, 23], as occurs with the hematopoietic stem cells (HSCs) [24, 25]. The pluripotent cells present in the stromal environment of the bone marrow appear to meet all these requirements, and are therefore, in analogy to the hematopoietic cell system, currently referred to as mesenchymal stem cells (MSCs) [16].

It must be noted that the term "marrow stromal cells," confusingly, has also been used to describe the adherent stromal monolayers in long-term bone marrow cultures or Dexter-type cultures, which can support hematopoiesis [26, 27]. In addition, selection methods, culture conditions and subsequent manipulations, may result in several distinct cell types, as evident from their different phenotypes, differentiation potential, and secretion products [28]. However, recent studies report that adult human bone marrow-derived stem cells cultured in an adherent monolayer are, if not practically indistinguishable, at least similar, both physically and functionally, independently of the method of isolation or proliferative expansion [28, 29].

MSCs are relatively easily isolated from the bone marrow and expanded *in vitro*. It was found that these stem cells play an important role in bone physiology, remodelling and hematopoiesis, and also may participate in the pathophysiology related to bone diseases.

Recently, these promising multipotential progenitor cells have been isolated from several other tissues, including cord blood, bone and adipose tissue, as described below. Most aspects of MSC biology, including their ontogeny, anatomical location in marrow, *in vivo* functions and transplantation biology remain to be elucidated. Attempts undertaken to reveal these issues have resulted in confounding results, principally due to the fact that different methods have been used to obtain and culture MSCs, to assess their differentiation potential, and to evaluate their capacity for self-renewal.

3. ISOLATION OF BM, UCB AND PB-DERIVED MSCS

3.1. BM-derived MSC

Dexter et al. were the first to propose the general idea that adult hematopoiesis could be propagated *in vitro* by creating a stromal microenvironment comparable to the microenvironment present in the bone marrow. This resulted in the controlled establishment of *in vitro* culture conditions for LT-BMC [26]. Subsequent studies demonstrated that adherent stromal cell layers, by creating favorable environmental conditions, could support maintenance and growth of hematopoietic stem cells (HSCs) *in vitro*. *In vivo*, the effects of the stromal compartment on the marrow microenvironment are mediated by both cell-cell interactions and the production of a vast array of growth factors and cytokines, which are necessary for normal hematopoietic function and differentiation [30]. Bone marrow stroma is currently the most accessible and therefore the most recurrent tissue source used to grow mesenchymal progenitors. Human MSCs are most frequently derived from aliquots of BM obtained from healthy donors undergoing marrow aspiration from the posterior iliac crest after informed consent and ethical approval [16, 17, 31]. The BM is usually collected in preservative-free heparin and BM mononuclear cells (BM-MNC) are obtained after density gradient centrifugation. In the majority of the reported conditions the light-density mononuclear cells are seeded at concentrations of $1–5 \times 10^6$ MNCs/cm^2 in relatively undefined media compositions, consisting of a basal medium (such as α-MEM or DMEM-LG) supplemented with selected batches of fetal calf serum (typically from 10% up to 20%) and/or other animal sera, thus impairing the study of physiologic signals required for efficient attachment and culture expansion [30]. Cells are commonly incubated at 37°C at 5% CO_2 in a humidified atmosphere. After allowing 24–48 hours for adherence to non-coated polystyrene, the supernatant is discarded and non-adherent cells are removed. The resulting population of cells serves as the primary *ex vivo* source of MSCs. Cells are expanded for approximately 14 days until subconfluence and can then be further processed by treatment with a Trypsin/EDTA solution after which cells are expanded through sequential passages to confluence. Most contaminating hematopoietic cells are progressively lost and after the second passage cultures contain a morphologically homogenous cell population. This is confirmed by FACS analysis, showing lack of expression of typical hematopoietic cell surface markers, such as CD45, CD34 and CD14. The use of varying culture conditions between the different laboratories, including addition of growth factors or cytokines and of certain batches of FCS in cultivating MSCs, may result in a moderately to significantly selective enrichment of stromal progenitor cells [28, 29].

3.2. UCB-derived MSC

Some controversy still exists as to whether MSCs can also be cultured from (mobilized) peripheral blood (PB) and umbilical cord blood (UCB). Mareschi et al. attempted to obtain MSCs from full-term UCB [32]. However, although both BM and UCB-derived MNCs were able to generate an adherent cell monolayer using the same culture conditions, cells obtained from UCB displayed the morphology and characteristics of hematopoietic cells and not those of MSCs. BM-MSCs expressed mRNAs for IL-6 and IL-11, which are known stimulators of hematopoiesis [28] and lacked expression of TGF-β1 and TNF-α, pleiotropic cytokines principally produced by monocytes and macrophages. In contrast, UCB derived adherent cells were not able to produce IL-11 mRNA as do BM-MSCs and measured expression of TGF-β1 and TNF-α was most probably due to the presence of monocytes-macrophages in the adherent layer. In addition, multi-nucleated cells found in the UCB adherent cultures were strongly positive for tartrate-resistant acid phosphatase (TRAP). This suggested that monocytes-macrophages present in these cultures might have spontaneously differentiated into osteoclasts. Furthermore, no signs of mesenchymal differentiation of the UCB-derived adherent cells were observed, because cells grown in specific culture medium died very quickly. Wexler et al. [33] used different culture conditions than Maresci et al., but obtained similar results. Although adherent cells were found both in BM-derived samples and full-term UCB samples, these cells were not similar, but in fact quite distinct. The adherent cells derived from UCB expressed a monocyte and macrophage phenotype, being CD45+ and CD14+. These cells produced a heterogeneous mixture of non-confluent cells, which could not be passaged. In concurrence with this, Ye et al. confirmed that the major cell types present in UCB monolayers were fibroblasts, macrophages and endothelial cells, lacking production of IL-11 [34]. The vast majority of the adherent cells derived from UCB and developed in standard Dexter-type long-term culture were also shown to be of hematopoietic origin and belonged to the monocyte-macrophage lineage [35]. More than 90% of the adherent cells expressed acid phosphatase and CD14, CD68 and CD115 (antigens expressed by macrophages) were expressed by 50%, 83% and 70% of the adherent cells, respectively. In keeping with these observations, IL-6 and TNF-α were both expressed. In contrast, no evidence for the presence of fibroblasts, osteoclasts or endothelial cells was found in these cultures, although CD1a, a dendritic cell marker, was expressed by a high proportion of the adherent cells in UCB-cultures (43–73%).

During fetal development, MSCs are circulating and can be isolated from unstimulated first-trimester fetal blood. However, the mean number of the obtained adherent colonies declines with advancing gestation [36]. Erices et al. evaluated the presence of UCB-derived cells with MSC characteristics [22]. They found that approximately 75% of umbilical cord blood harvests gave rise to cultures of adherent cells, which consisted predominantly of osteoclast-like cells (OLC). The remaining 25% of the cultures contained primarily mesenchymal like cells (MLC) [22]. OLCs displayed the morphology and characteristics of multinucleated osteoclasts, expressing several markers of osteoclasts, such as a strong TRAP activity and expression of antigens CD45 and CD51/61 (vitronectin receptor) [37]. In contrast, MLCs displayed a fibroblast-like morphology and expressed several MSC-related antigens, but did not express endothelial or myeloid antigens. Additionally, MLCs could differentiate into several mesenchymal lineages when cultured under appropriate conditions. OLCs were especially dominant in full-term UCB, whereas MLCs were mainly found in pre-term UCB collections. Moreover, the content of MLCs in UCB was inversely proportional to fetal age, a trend also previously observed for hematopoietic progenitors. Taken together, these data suggested that, similar to hematopoietic stem cells, mesenchymal progenitors and stem cells travel during the early fetal development most likely via the cord blood, from their original niche in the fetal hematopoietic sites, such as the liver, into the bone marrow [22, 38]. This would explain the discrepancies with previous studies, in which primarily full-term harvests of UCB were utilized.

Interestingly, MSC-like cells were recently obtained from the subendothelial layer of the human umbilical cord blood vein [39]. These cells did not express monocyte/macrophage antigens, were α-smooth muscle actin positive, and deposited fibronectin and type I collagen. In addition, these cells terminally differentiated *in vitro* in at least the adipocytic and osteogenic lineages.

3.3. PB-derived MSC

Bone marrow and granulocyte colony-stimulating factor (G-CSF) stimulated peripheral blood samples were collected and used for culture-expansion. Although MSCs were routinely detected from central samples of bone marrow using *in vitro* cultures and their differentiation potential confirmed by the ceramic cube assay, which showed *in vivo* bone formation, these cells were not obtained from any of the PB collections (both patients and their healthy donors) [40]. Furthermore, no adherent stromal cell layer could be obtained after culture of light-density mononuclear cells from peripheral blood leukapheresis collections of both patients with solid tumors and hematological malignancies and healthy donors, showing that in humans MSCs present in the bone marrow are not mobilized towards peripheral blood by chemotherapy and/or growth factor stimulation [41]. In comparison, cells with characteristics of mesenchymal progenitors were detected in growth factor-mobilized peripheral blood stem cell collections, harvested from patients with breast cancer [21]. However, these cells also shared characteristics with monocytes and pre-osteoclasts and appeared similar to the cells described by Purton and colleagues, who reported the presence of adherent pre-osteoclastic cells in G-CSF mobilized peripheral blood from healthy donors [42].

Zvaifler and colleagues collected unstimulated PB from over a hundred healthy individuals and found in a fraction of elutriated blood a subset of CD34 negative plastic adherent cells, which displayed both morphological and phenotypical characteristics of MSCs, such as a strong positive staining for SH2 (endoglin) and STRO-1, and expression of SDF-1 [43]. Furthermore, these PB-derived MSC-like cells were able to differentiate under appropriate culture conditions into fibroblasts, osteoblasts and adipocytes. However, the same elutriation fraction also contained a large amount of monocytes (65%), of which a subset spontaneously formed multinucleated osteoclast-like cells.

In the murine model, CFU-F circulate in unstimulated peripheral blood and represent a stromal cell population that can migrate into hematopoietic organs. Phenylhydrazine treatment of mice resulted in a threefold increase in blood CFU-F numbers [44]. In addition, Kuznetsov and colleagues reported the presence of adherent cells, isolated from the unstimulated blood of four mammalian species, including mice, rabbits, guinea pigs (cardiac puncture) and humans (venous blood) [45]. However, there was significant variation in colony-forming efficiency (CFE, *vide infra*) across animal species and between individual donors. Cultures of all species except humans (only two colonies were found) contained cells with two different types of morphology. Most colonies consisted of fibroblast-like cells, but a number of colonies contained cells with a more polygonal morphology. The phenotype of both groups of cells was nearly identical between all species and closely resembled the profile of their respective BM-MSCs. However, in contrast to human BM-MSCs, which express the putative MSC markers STRO-1, endoglin and MUC-18, the human PB-derived adherent cells were negative for these specific surface markers (see Table 1.2) [46]. Furthermore, BM-derived MSCs commonly express Alkaline Phosphatase, whereas PB-derived cells of all four species were consistently negative for expression of this enzyme. The ceramic cube assay revealed bone formation in up to 50% of the strains implanted, depending on the species, whereas adipogenic differentiation was confirmed for cells of all four species by *in vitro* differentiation assays.

These studies suggest that both species and culture-related conditions are of great importance and may have resulted in contradictory observations.

4. TISSUE SOURCES OF MSCS

4.1. Adipose-derived MSC

Recently, adipose tissue has been found to contain a population of stromal cells with multilineage potential [47, 48, 49]. Because adipose or fatty tissue is, similar to bone marrow, ontogenetically derived from the embryonic mesoderm, it is not surprising that the differentiation potential and phenotype of these progenitor cells resembles that of BM-derived MSCs. Fatty tissue is build up out of a heterogeneous stromal cell population, which can be relatively easy separated by collagenase treatment of the extracellular matrix and subsequently propagated in culture. The evolving population of cells display a fibroblast-like morphology, and can be expanded extensively in culture while maintaining a relatively stable population doubling rate. Staining for βGal revealed that the percentage of senescent cells increased from undetectable at passage 1 to approximately 15% at the time of passage 15 [47]. These plastic-adherent cells, which resemble BM-MSCs and are therefore referred to as adipose-derived MSCs (AMSC), are of mesenchymal origin, as shown by immunophenotyping. AMSC derived after processing of lipoaspirates as a source of fatty tissue are also referred to as processed lipoaspirate (PLA) [47]. Initially, PLA or AMSC cultures may contain contaminating cells, which mainly consist of fibroblasts, pericytes, mast cells, endothelial cells, and smooth muscle cells. Other possible sources of AMSC are abdominal subcutaneous adipose tissue and mesenteric fatty tissue.

Similar to BM-derived MSCs, AMSC derived from a number of species, including human, rat and rabbit, can differentiate *in vitro* into adipogenic, chondrogenic, myogenic (both skeletal and heart muscle), osteogenic and even neuronal like-cells [48, 49, 51]. Furthermore, AMSC express multiple CD marker antigens corresponding to those observed on BM-MSCs [50, 51]. Although both stromal cell populations originate from different tissue sources, they have a very similar expression pattern of adhesion and receptor molecules. While progressing from a progenitor phenotype towards committed and more restricted cells, AMSC express a spectrum of lineage-specific genes and proteins.

Although AMSC possess most of the typical characteristics of BM-MSCs, they do also display unique features, including differences in expression of surface markers, time and/or type of gene expression and response to differentiation media. For example, AMSC are positive for CD49d and negative for CD106 and STRO-1, whereas the exact opposite is reported for BM-MSCs [50, 51]. Furthermore, in contrast to BM-MSCs, AMSC do not express BMP-2 and dlx5, which are known regulators of osteogenic genes. Although AMSC could be directed to differentiate into several lineages, no chondrogenic differentiation was obtained in BM-MSC cultures with the same induction media. In addition, AMSC were more sensitive to osteogenic induction by 1,25 dihydroxyvitamin D3, whereas BM-derived MSC showed a better response after treatment with dexamethasone. The importance and consequences of these differences are currently being explored.

Adipose tissue is relatively easily accessible and AMSC can be obtained in large numbers, cultivated and differentiated into both mesenchymal and ectodermal lineages. Therefore, in addition to BM-MSC, AMSC represent another multipotential population of mesenchymal progenitor cells [51].

4.2. Bone-derived MSC

Adult human osteoblastic cells (hOB) can be derived from explant cultures of trabecular bone and are capable of at least trilineage differentiation. When cultured with differentiation inducing media and under specific culture conditions, hOB were able, similar to BM-MSCs, to differentiate *in vitro* into adipose cells, osteogenic cells (both monolayer cultures) and chondrocytes (high-density pellet culture) [52]. Additional evidence of the pluripotentiality of bone-derived cell populations comes from studies using specific cell populations. Fetal rat calvariae were subjected to an enzymatic digestion, after which the cells were cultured for 2 days [53]. The adherent stromal cells were detached and sorted by flow cytometry for particle size (forward scatter) and cytoplasmatic granularity (side scatter). A population of small, slowly cycling cells with low cytoplasmic granularity, which were termed S cells, displayed several features characteristic of (mesenchymal) stem cells, such as self-renewal, large expansion potential and the ability to generate a spectrum of descendents ranging from restricted to terminally differentiated daughter cells. S cells were found to be able to differentiate *in vitro* into a number of mesenchymal lineages, including cartilage cells, adipocytes, smooth muscle and bone. In addition, S cells were able to support hematopoiesis when co-cultured with BM-derived nonadherent hematopoietic cells and functioned as competent stromal cells. In addition, Grigoriadis and colleagues isolated a multipotential clonally derived cell population with from fetal rat calvariae, termed RCJ 3.1 [54]. This clonal cell line differentiated in a time-dependent manner in presence of ascorbic acid, sodium beta-glycerophosphate and dexamethasone, into muscle cells (observed at days 9–10), adipocytes (day 12), chondrocytes (day 16) and bone (day 21). Gronthos et al. sorted four different cellular subsets, obtained from primary cultures of normal human trabecular bone, on the basis of the expression of the putative stromal precursor cell marker STRO-1 and the osteoblastic marker alkaline phosphatase (ALP) [55]. The STRO-1+/ALP-phenotype corresponded with the earliest subset of cells and could give rise to all the four subpopulations. Compared to the other phenotypes, these cells displayed a decreased capacity to form mineralized bone nodules and did not express the bone-related markers bone sialoprotein, osteopontin or parathyroid hormone receptor. The bulk of the bone-derived cells displayed either a STRO-1-/ALP+ or a STRO-1-/ALP-phenotype, representing each a group of terminally differentiated osteoblasts. The last subset, comprising a STRO-1+/ALP+ population, contained a group of pre-osteoblastic cells, which were not yet fully differentiated. Thus, bone-derived cell cultures can be obtained by distinct experimental approaches and from different species and contain both undifferentiated mesenchymal progenitor cells, transit amplifying cells (intermediate stage of differentiation) and committed osteogenic precursors.

4.3. Muscle-derived MSC

After enzymatic disaggregation of human, avian, rodent, and rabbit skeletal muscle tissue, the isolated cells can be propagated in culture, giving rise to an adherent stromal cell culture [56]. In cultures grown in medium supplemented with special batches of horse serum, the stellate cells maintained an undifferentiated phenotype, but in the presence of dexamethasone they differentiated and acquired the phenotypical and morphological appearance of skeletal and smooth muscles, bone, cartilaginous, or adipose tissue. These results demonstrate the presence of mesenchymal progenitors in skeletal muscle [56]. In irradiated, immunodeficient mice, transplantation of radioactive labelled, cultured muscle precursor cells, demonstrated that virtually all newly donor-derived muscle is produced by a relatively small subset of muscle-resident cells that divide slowly in culture, but become activated and proliferate more rapidly after en-grafting [57]. Presumably, these cells represent a distinct population of uncommitted MSCs, which seems different from muscle satellite cells, generally considered to be the muscle stem cell [58].

In addition, heart muscle also seems to contain a population of MSCs. Adherent stromal cell cultures derived from the myocardium of 3–5 day old rats, contained a population of cells with a stellate morphology. Dexamethasone induced differentiation, resulted in the appearance of adipocytes, osteoblasts, chondrocytes, smooth muscle cells, skeletal muscle myotubules, and cardiomyocytes [59].

4.4. Cartilage-derived MSC

A proportion of culture-expanded human articular chondrocytes demonstrated at least trilineage differentiation potential, including chondrocytic, adipogenic, and osteogenic potential. However, these cells were distinct from the BM-derived MSCs in that the cells, when loaded into porous calcium-phosphate ceramic cubes only formed cartilage and not bone [60]. Therefore these cells may represent a subpopulation of mesenchymal progenitor cell rather than true stem cells.

4.5. Other sources of MSC

Currently, other possible sources of MSCs are being explored. These include tendon [61], vascular tissue [62, 63], dental pulp [64] and a variety of fetal tissues [36, 65].

5. IN VITRO ASSAY OF MSCS: THE COLONY FORMING UNIT-FIBROBLAST (CFU-F)

The bone marrow stroma, which consists of a loose network of reticular fibers and cells, can be relatively easy manipulated and processed to separate the stroma from the hematopoietic cells. The resulting single cell suspension is then subjected to a density centrifugation, after which the mononuclear cells are recovered from the buffy coat. Alternatively, a number of distinct methods can be utilized for further enrichment of the stem cells, such as magnetic-activated cell sorting (MACS) or fluorescent activated cell sorting (FACS). The obtained cells are then plated at low density in culture medium supplemented with fetal calf serum (FCS). The bone marrow-derived stromal cells can be easily separated from the non-adherent hematopoietic cells by their ability to adhere to plastic surfaces: The supernatant containing the non-adherent cells is discarded after an initial incubation, while the mesenchymal stem cells remain attached to the culture vessel. After a series of repeated washings with a buffered salt solution, contamination of hematopoietic cells in the cultures is reduced to a minimum. Under appropriate culture conditions, distinct colonies of fibroblast-like cells are formed, each of which is presumably derived from a single precursor cell. These colonies are now commonly referred to as the colony forming unit-fibroblast or CFU-F [12, 46].

Originally, the progressive studies, performed by Friedenstein and colleagues, resulted in isolation of CFU-F colonies from rodent BM [12]. Subsequent efforts of Owen et al, showed that CFU-F colonies derived from the BM of basically all species, including humans, are heterogeneous in size, morphology and potential for differentiation [66], prompting the suggestion that they are derived from clonogenic progenitors at various stages of commitment. A proportion of the colonies is large in size and demonstrates extensive replating potential after passaging [67]. The high proliferative potential of these clonogenic colonies led Friedenstein to propose that within the CFU-F compartment, a small group of precursor cells with stem cell characteristics are present. In contrast, the smaller colonies with less proliferative and differentiation capacity were considered to be more restricted mesenchymal precursor cells. Using *in vitro* assays to determine the differentiation profile of individual colonies, several groups demonstrated subpopulations within the mesenchymal stromal population [68], including cells with osteogenic potential only or those also displaying chondrogenic or adipogenic lineage potential. Cells derived from individual colonies have also been tested *in vivo* using implanted diffusion chambers in a rat model [67]. These results were similar to those from *in vitro* differentiation studies: although full mesenchymal differentiation potential was demonstrated by some colonies, others showed a restricted lineage potential. These observations demonstrated that the cloned populations derived from an individual CFU-F contain both MSCs and precursors with a more restricted differentiation potential.

The frequency of CFU-F within the MNC subset can be determined by the colony-forming efficiency (CFE) [69]. The CFE is highly variable, depending a great deal on the batches of FCS used to define the culture conditions, and additional requirements for optimal CFE differ significantly between distinct species. For example, CFU-F obtained from rodents are more difficult to maintain in culture and irradiated marrow feeder cells or addition of external growth factors, such as Leukemia Inhibitory Factor (LIF), are an absolute necessity in order to obtain the maximum number of CFU-F (100% CFE). In contrast, human cells are for optimal CFE feeder cell-independent [15]. The differences in the need for and the effect of specific growth factors for optimal growth may result from differences between species in the levels of the corresponding receptors on the MSCs [15]. In addition, other factors may influence the quality of the obtained samples and consequently the CFE, such as the procedure to harvest the marrow [70, 71, 72], the low frequency of MSCs in marrow harvests (typically around 2 to 5 MSCs per 1×10^6 MNCs) [73], the method of selection used to obtain and maintain the MSCs in culture (for example, positive selection or negative depletion on the basis of phenotype), and the age or condition of the donor from whom the MSCs were prepared [72, 74, 75]. The growth factors or cytokines, required to stimulate formation of CFU-F are not completely elucidated, but do at least include platelet-derived growth factor-BB (PDGF-BB), epidermal growth factor (EGF), basic fibroblast growth factor (bFGF), transforming growth factor β (TGF-β), and insulin-like growth factor (IGF) [14, 76, 77, 78] and will be discussed in the following paragraph.

6. EXPANSION OF MSC IN SERUM-DEPRIVED CULTURES

Serum-deprived culture conditions have been and are continuously being developed by several groups to enhance the understanding of and to gain insight in the effects of certain cytokines, growth factors, hormones and other additives on the colony growth of MSCs. Serum plays an essential role in the support of cell growth *in vitro* by providing critical factors, such as amino acids, minerals, transport proteins carrying lipids, growth factors and hormones, by supplying vitamins and attachment factors, by acting as a pH buffer and by providing protease inhibitors [78, 79]. However, serum may also act as a source of potentially cytotoxic factors, such as pyrogens. Therefore, supplementation of media with a large volume of FCS (typically up to 10–20%) [72, 80] can either mask or modify the response of MSCs to exogenously added factors. Furthermore, there is a high degree of variability between different batches of serum, which results in a relative lack of characterization of serum components. The use of different lots of FCS therefore requires extensive testing of serum, and makes the development of a chemically defined medium a necessity [78].

To promote attachment of MSC under serum-deprived conditions, dishes can be precoated with a fibronectin in PBS (10 μg/cm^2) for 2 hours at room temperature. Just before the cells are plated, excess of fibronectin is then removed by rinsing once with PBS [77]. Gronthos and Simmons cultured STRO-1$^+$ cells, sorted from human BMMNC in α-MEM supplemented with 10 μg/ml insulin, 2% BSA, 80 μg/ml LDL, 200 μg/ml iron saturated transferrin, 2 mM L-glutamine, sodium phosphate, 5×10^{-5} M β-mercapto-ethanol and 100 U/ml penicillin and 100 μg/ml streptomycin with or without supplementation of additional growth factors [77, 81]. Serum-deprived medium, supplemented with 10 ng/ml of EGF, PDGF, IGF-1 and bFGF, did not stimulate development of CFU-F. However, if 100 μM L-ascorbic acid-2-phosphate (ASC-2P) and 10^{-8} M dexamethasone (Dexa) were added to the medium, CFE was similar to control cultures supplemented with 20% FCS. When ASC-2P or Dexa were supplemented without the addition of exogenous growth factors, no CFU-F growth occurred [15]. Although cultures, supplemented with PDGF and EGF, in presence of ASC-2P and Dexa, showed an equivalent ability to support CFU-F growth compared to control cultures, the average colony size of the CFU-F was significantly increased. In contrast, first passage MSCs derived from rat BM, obtained after initial culturing under serum-replete conditions, showed optimal growth when cultured in a serum-free medium, which consisted of 5 μg/ml insulin, 0.1% LA-BSA, 10 ng/ml PDGF-BB, 1 ng/ml bFGF in a base medium of 60% DMEM-LG with 40% MCDB-201 [78]. Additionally, mouse and human stromal cell cultures were stimulated by a serum-free conditioned medium (SF-CM), consisting of α-MEM supplemented with 2mM L-glutamine, 100 U/ml penicillin, 100 μg/ml streptomycin sulphate, 10^{-8} M dexamethasone, 10^{-4} M L-ascorbic acid phosphate magnesium salt n-hydrate and 0.5% ITS+, conditioned by marrow cells. It was found that neutralizing antibodies against PDGF, TGF-β, bFGF and EGF specifically, were able to suppress all colony formation. However, growth factor dependence varied between the different species, and the most profound inhibition of mouse CFU-F formation was induced by anti-PDGF, anti-bFGF and anti-EGF, whereas in human cultures anti-PDGF and anti-TGF-β were most effective [15].

Differences between the above mentioned studies are in part due to the use of distinct methods of selection of the cells and the culture conditions (with or without feeder layer or conditioned medium). Of some importance may be the use of high concentrations of growth factors or addition of BSA that are not recombinant and may therefore contain traces of multiple serum activities (for example, growth factor-binding proteins).

7. DIFFERENTIATION POTENTIAL AND PROLIFERATIVE HIERARCHY

7.1. Orthodox differentiation potential

The differentiation potential of stromal cells derived from a number of different mesoderm-derived tissues has been considered for a prolonged time to be restricted exclusively to mesenchymal lineages. The stem cell-like nature and characteristics of MSCs are now generally accepted and these cells are now commonly regarded as a second class of adult stem cells, in addition to HSCs, that populate the BM. However, they also constitute a variety of other adult tissues, as described above. In *in vitro* culture conditions, MSCs maintain their self-renewal capacity for extended periods and have the ability to generate a large quantity of different mesenchymal cell types. As such, these cells take part in the regeneration of mesenchymal tissues, in response to injury. Throughout a variety of species MSCs were found to be able to differentiate into several tissues including bone [82, 83], cartilage [84], stroma [14], adipose [11], tendon [85], but also neural tissue [86, 87, 88, 89], smooth muscle [90], and cardiac muscle [91, 92, 93, 94] and a variety of other connective tissues [14], as discussed below. The differentiation potential of MSCs into the "standard" mesenchymal lineages is also known as the orthodox differentiation.

Only a small proportion of the initial adherent BM-derived stromal cells are actually pluripotent and capable of multilineage differentiation, as demonstrated by lineage specific *in vitro* assays. Human MSCs maintain their osteogenic potential for up to approximately 40 doublings in culture, even after cryopreservation [70, 95]. However, uncharacteristic of "true" stem cells, MSCs display a finite lifespan and also their multilineage differentiation potential is not unlimited, as verified by Muraglia et al. [96]. Clones derived from non-immortalized human BM-derived MSCs exhibited at least trilineage potential, as shown by their *in vitro* differentiation into osteo-, chondro- and adipogenic cells. At increasing cell doublings, clones firstly lost their adipogenic and subsequently also their chondrogenic potential, resulting in the generation of cells with restricted osteochondrogenic and eventually only osteogenic differentiation potential. When cultured under optimal conditions in medium supplemented with FGF-2, approximately one third of the clones, displayed trilineage potential and therefore contained a subset of early mesenchymal progenitors. After subculturing, these cells eventually approached senescence and progressively lost their multilineage potential, giving rise to bi- and ultimately monopotential cells only, suggesting the presence of a possible model of predetermined differentiation of the BM-MSCs [96]. These results were confirmed by DiGirolamo et al. [72], who demonstrated that cells after extensive subculturing entered senescence, as evident by their progressive loss of the ability to differentiate into the adipocytic lineage, while maintaining their osteogenic potential. Only a small selected group of cells retained their multipotentiality through a number of passages, whereas others either entered differentiation or began to senesce. In contrast to human MSCs, thus far no hierarchical program that completely covers the description of differentiation potentials of mouse MSCs has been reported [108].

In addition, the cellular population, which can be isolated from BM-MNC on the basis of STRO-1 expression, contains pluripotent progenitor cells, which are capable of *in vitro* differentiation into four distinct lineages, including hematopoiesis-supportive stromal cells with a vascular smooth muscle-like phenotype, adipocytes, osteoblasts and chondrocytes [97, 98, 99]. Furthermore, STRO-1 positive cells remain pluripotent through multiple rounds of subculturing [99]. These data therefore suggest that STRO-1 can be used to select a subset of mesenchymal progenitor cells, which are either similar or identical to the MSCs, isolated by their plastic-adherence.

However, the inevitable and eventual loss of the multipotentiality as the MSCs are replicated in culture may have important implications for the employment of cultivated MSCs. The use of these cells in future cell therapy or after genetic engineering for clinical purposes may be seriously limited if these cells can not retain their pluripotentiality after culture [16, 72, 87, 100]. Moreover, these results altogether imply that the assessment of the differentiation potential of MSCs, that have been isolated and maintained in culture for extended periods, is extremely important and should be confirmed by at least two distinct differentiation assays [77, 97]. Unfortunately, even if the expression of several phenotype related genes and differentiation related markers can be demonstrated, and even if the production of, for example, mineralized bone nodules or lipid vacuoles are proven by histo- or immunochemical staining, these findings merely represent the behaviour of cells which are removed from their physiological environment and therefore do not necessarily mimic the behaviour or pluripotentiality, which they may or might have displayed *in vivo* [6]. Although attempts have been undertaken, it remains virtually impossible to predict which one of the cells, present within a certain stromal cell population, will be or behave as a true "stem cell" and the identification of those cells can therefore only be performed retrospectively by employing an appropriate assay [46]. The behavioural difference of MSCs between *in vitro* and *in vivo* chondrogenic potential demonstrate this conflict clearly. The occurrence of selectively chondrogenesis in transplantation assays is a rare phenomenon, whereas it is readily induced *in vitro*. However, optimal induction of chondrogenesis in cultures requires the presence of at least Dexamethasone and TGF-β1, TGF-β2 or TGF-β3 in addition to the presence of a closed system, such as a 3-dimensional aggregate (or pellet) culture, which imitates the *in vivo* precartilage condensation [84], and in which low oxygen tensions facilitate the induction of chondrogenesis [6, 101]. Therefore, the actual differentiation potential of selected and cultivated cells depends heavily on the culture conditions employed in *in vitro* assays and on environmental influences in the case of transplantation studies [46]. For an overview of the differentiation potential and related factors, see Table 1.1.

7.2. Unorthodox differentiation potential and stem cell plasticity

Until recently it was considered that stem cells in adults could give rise only to tissues where the cells are resident. However, newly presented facts necessitated an alteration of, or more specifically, and addition to this concept. Recent studies have implied that adult tissue-derived progenitor cells may exhibit the "unorthodox" potential to differentiate into unrelated tissues, in addition to tissues derived from the same embryonic germ layer. Hematopoietic stem cells capable of differentiation into all cell elements of the blood, can also be a source of liver cells, such as hepatic oval cells [102], hepatocytes [103] and cholangiocytes [104], and can be induced to differentiate into neural cells [105, 106]. In return, neural stem and progenitor cells served as a source of both early and restricted hematopoietic precursors after engraftment into the hematopoietic system of irradiated mice [107].

This unexpected differentiation potential appears also to be the fact for MSCs (Table 1.1, Figure 1.1). The differentiation pathways of MSCs are not as irreversible as previously assumed, since even morphologically fully differentiated cells from a certain lineage may reverse from their original destination and return to a more early stage of differentiation and may, when cultured in presence of specific induction factors, express characteristics of intermediate phenotypes while exchanging one particular phenotype for another [89, 108]. Even more, several groups have now irrefutably demonstrated that MSCs exhibit this unorthodox differentiation potential and these cells therefore appear to belong characteristically to the family of putative somatic stem cells [46, 86, 87, 89, 109]. Adult MSCs can be obtained from several tissues that display transgermal plasticity (i.e. the ability to differentiate into phenotypically unrelated to cells) and are a likely candidate for the vacant position of common precursor cell of all adult stem cells. Hence, in addition to their ability for self-renewal and extensive generation of

Table 1.1. Minimal stimuli necessary for induction of differentiation of bone marrow-derived MSCs in vitro, and the molecular and cellular markers involved*.

Cell type	Stimuli	Molecular markers	Cellular markers	References
Adipocytes	Dexa + IBMX + IND + Insulin	PPARγ2, aP2, LPL, C/EBP-α/β, retinoids	Cytoplasmic lipid vacuole formation	11, 17, 22, 108, 113
Chondrocytes	Dexa + ASC-2P + TGFβ3/β1 + ITS + ± BMP-6	Cbfa-1, collagen type II and IX, Sox-5, -6, -9, aggrecan	Proteoglycans, collagen type II and X	11, 84, 96, 108
Osteoblasts	Dexa + ASC-2P + βGP	Cbfa-1, bone sialoprotein, osteopontin, osteocalcin, ALP, PTH-R	Mineralized matrix, bone nodules	11, 22, 55, 72, 108
Tenocytes	BMP-12	Collagen type II, proteoglycans	Improved bio-mechanism of implanted tendon	85, 190
Hematopoietic supporting stroma	HC + HS	MEF-2C, BTEB-2	Maintenance and support of HSC	28, 108
Skeletal muscle cells	5-AZ, Amphotericin B, HS + HC	MyoD, Myf 5 and 6, MEF-2, myogenin, MRF4, myosin	Multinucleated, contractile cells	116
Smooth muscle cells	PDGF-BB, TGFβ	ASMA, EDa/bFN, metavinculin, h-caldesmon, SM myosin, TSP-1, desmin	Single nucleated, fusiform cells	90, 108
Cardiac muscle cells	5-AZ	β 1-, β2-adrenergic, M1, M2-muscarinic receptors, ANP, BNP, desmin, β-myosin heavy chain, α-actinin, cardiac troponin T, phospholamban	Spontaneously beating cells, sarcomeric organisation of contractile proteins, incorporation in myocardium	91, 93, 94, 117, 118, 119

(Cont.)

Table 1.1. (Cont.)

Cell type	Stimuli	Molecular markers	Cellular markers	References
Astrocytes	Dexa + DMSO	GFAP	Engraftment into neonatal brain	86, 87, 114
Oligodendrocytes	Local cues	GalC, O4	Myelin production, oligodendrocyte morphology	115
Neurons	BHA + bFGF + forskolin + DMSO + heparin + K252a + KCl, VA + N2 supplement + PDGF	β-III tubulin, tau, NF-M, TOAD-64, synaptophysin, NSE, ChAT	Neuronal morphology	89, 109, 114, 115
Schwann cells	βME + RA + forskolin + bFGF + PDGF + Heregulin	p-75 (NGF-R), S-100, GFAP, O4	Myelination of regenerated nerve fibers, support of nerve fiber regrowth	88

(Abbreviations: Dexa: dexamethasone; IBMX: Isobutylmethylxanthine; IND: Indomethacin; PPARγ2: peroxisome proliferation-activated receptor γ2; ASC-2P: Ascorbic acid-2-Phosphate; βGP: beta-Glycerophosphate; BMP: Bone Morphogenetic Protein; HC: Hydrocortisone; HS: Horse serum; 5-AZ; ASMA: alpha smooth muscle actin; EDaFN: FN isoform ED a and b; TSP-1: Thrombospondin-1; GFAP: glial fibrillary acidic protein; BHA: butylated hydroxy anisole; VA: Valproic acid; βME: beta-Mercapto ethanol; RA: Retinoic acid; bFGF: basic-Fibroblast growth factor; PDGF: Platelet-derived growth factor; NF-M: Neurofilament-M; NSE: neuronal specific protein; ChAT: Choline acetyltransferase.)
Adapted from reference 31.

a broad spectrum of mesenchymal daughter cells, MSCs are distinguished from other progenitor cells by their intriguing plasticity for non-mesenchymal lineages.

Enver and colleagues showed that hematopoietic stem cells simultaneously express a variety of genes characteristic of multiple distinct hematopoietic lineages prior to commitment to a particular lineage [110, 111]. They concluded that differentiation of stem cells in general appeared to be an intrinsic process, but as the cells are progressing towards a differentiated phenotype, extrinsic influences become more important. In analogy to this, MSCs, isolated from Dexter-type cell cultures up to a purity of more 95%, were found to coexpress genes specific for a variety of mesenchymal lineages, including adipocytes, osteoblasts, fibroblasts, and muscle [112]. No evidence was found for the presence of each of these cells in the undifferentiated primary cell culture, indicating that the expression of the specific lineage markers represented the activity of a single cell population [112]. In addition, Woodbury et al. demonstrated

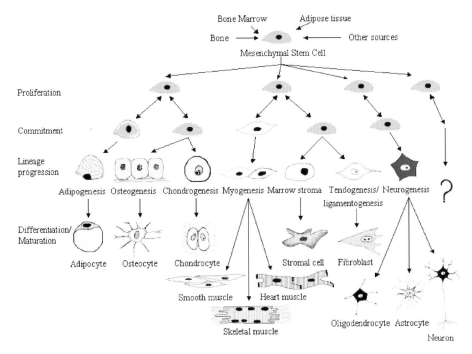

Figure 1.1. A schematic representation of MSC plasticity, depicting the currently known stepwise transitions from the mesenchymal stem cell into terminally differentiated phenotypes. This picture is incomplete and does not show all potential transitions, nor does it display the intermediate phenotypes, which occur when MSCs move between different lineages.

that previously untreated MSCs expressed the prototypical mesodermal genes, such as SM22α, myosin and leptin, in addition to germ line specific genes (protamine 2), endodermal genes (ceruloplasmin) and ectodermal genes (NMDA glutamate binding protein, APP, syntaxin 13 and brain-specific aldolase C) [89]. These results suggested that instead of being regarded as classically "undifferentiated," MSCs should be considered to be initially "multidifferentiated" cells [89, 112]. Furthermore, the presence of a diversity of ready accessible genes, specific for a variety of lineages, allows and facilitates the rapid response of MSCs in reaction to differentiation induction. However, the coexpression of multiple lineage-specific genes does not necessarily imply that MSCs will actually transcribe these genes into active products, nor does it represent proof for the multipotentiality of these cells.

Additionally, it was found that MSCs could be induced to overcome their mesenchymal fate and differentiate after treatment with a relatively simple, fully defined medium, into cells with neuronal characteristics within 5 hours of induction [89, 109]. Strikingly, this differentiation was reversible: After incubation of the cells in neuronal induction medium for up to 24 hours and reculturing of the cells in standard growth medium, after initial conversion to neuronal phenotype, cells readapted an MSC-like phenotype.

When human MSCs were cultured in presence of adipogenic agonists, over 90% of the cells differentiated into functional adipocytes within 6 days [113], as shown by increased levels of peroxisome proliferation-activated receptor $\gamma 2$ (PPAR$\gamma 2$) mRNA in the early stages of differentiation, lipoprotein lipase (LPL) mRNA at somewhat later stages, and formation of intracellular lipid droplets. When kept in maintenance medium, these cells excreted the lipid clusters and dedifferentiated into fibroblast-like stem cells, while PPAR$\gamma 2$ and LPL mRNA expression diminished. From these studies it becomes increasingly clear that fate determination and differentiation are not as irreversible and straight forward as previously assumed, and in presence of appropriate external or local clues, stem cells may differentiate and even redifferentiate into the same lineage or possibly into other unrelated lineages.

In addition to the neuronal differentiation described above, *in vivo* transplantation studies of MSCs have demonstrated engraftment and functional differentiation in other ectodermal cells, such as astrocytes, oligodendrocytes and Schwann cells [86, 87, 88, 114, 115].

Multinucleated myotubes appear after *in vitro* growth of rat BM-derived MSCs in the presence of 5-azacytidine [116]. Alternatively, myogenic differentiation can be induced by treatment of AMSCs with a mixture of FCS and horse serum supplemented with dexamethasone and hydrocortisone [51]. Intra muscular injection of 10^6 genetically marked unfractionated BM cells per muscle into immunodeficient *scid/bg* mice revealed that marrow-derived cells could functionally engraft into areas of induced muscle injury, where they demonstrated myogenic differentiation, and augmented the regeneration of the damaged muscle fibers [90]. In a second experiment, the BM-derived cells were separated into two fractions—adherent and non-adherent cells—and implanted directly into regenerating mouse muscle. Both fractions were found to be capable of contributing to muscle formation. In addition, to test whether BM could be recruited to sites of muscle injury, unfractionated BM cells were transplanted into the peripheral circulation of recipient mice, and muscle regeneration by BM-derived cells was confirmed. However, the actual contribution was limited and muscle-derived satellite cells, which were used as controls and injected at a dose of 5×10^5 cells/muscle, contributed in larger quantities to the regenerating muscle [90]. Additionally, several groups have shown that BM-derived MSCs are able to differentiate into functional cardiomyocytes, both *in vitro* and *in vivo* [91, 93, 94, 117, 118, 119].

Recently, Reyes et al. reported the presence of a mesodermal progenitor cell (MPC) in human bone marrow, selected by negative depletion of Glycophorin A and CD45, and cultured in presence of EGF and PDGF-BB. MPC could be expanded extensively and isolated single cells were able to give rise to endothelial cells (visceral mesoderm), when stimulated with VEGF, in addition to the mesenchymal lineages previously described for MSCs (limb-bud mesoderm) [120]. These MPCs were able to generate each of the cellular components, which are present in the healthy bone marrow microenvironment, including adipocytic, osteogenic, endothelial and hematopoiesis-supporting stromal cells. Jiang et al. further demonstrated the presence of multipotential adult progenitor cells or MAPC in rodent BM. These cells could be expanded for more than 80 population doublings without evident senescence and differentiated *in vitro*, at a single cell level, as confirmed by retroviral marking, into MSCs, visceral mesoderm (more than 90% positive for CD31, CD62E and von Willebrand Factor), neuroecto-derm (more than 90% positive for either GFAP, GalC or NF-200) and endodermal cells (60% stained positive for albumin, CK18 or HNF-1). When microinjected into early blastocysts, MAPC derivatives were found in most resulting somatic cell types. On transplantation of these cells into a non-irradiated host, MAPCs engrafted and differ-entiated into the hematopoietic tissues and liver, lung and gut epithelial cells. No MAPC engraftment was observed in heart and skeletal muscle, kidney, skin or brain. However, since MAPCs could be expanded extensively without any obvious signs of senescence and with maintenance of their pluripotentiality, these cells represent a potential source for gene therapy [121].

8. CHARACTERISTICS OF HUMAN BM-DERIVED MSCS

8.1. Morphology

MSCs, which can be isolated from other cells by their adherence to plastic, display a range of characteristics, that have been described in detail by several different groups [16, 67, 68, 100]. However, this has resulted in the development of a variety of distinct, but closely related culture protocols, in order to obtain the highest possible purity of MSCs. Unfortunately, the fact that different groups use different stromal tissues and different species as a source of MSCs, which are subsequently selected from those particular tissues under different conditions, has only increased the confusion. For example, although primary cultures of human BM-derived MSCs initially contain some hematopoietic cells, these cells are lost within two or three passages. In contrast, cultures of murine BM-MSCs contain, in general, a relatively large quantity of contaminating lymphohematopoietic cells (CD11b and CD34 positive) and the proportion of MSCs of some inbred strains was so low that standard methods were not sufficient to obtain high enough yields of a relatively homogenous population of MSCs [72, 191].

On morphological examination of the BM-derived CFU-F, the heterogeneity of the colonies becomes immediately evident. The colonies differ both with respect to size, as a result of the variability in doubling time, and cell morphology (large, flat cells or more spindle-shaped). Kuznetsov et al. [14] and Satomura et al. [76] demonstrated that single-cell derived colonies, obtained by plating early passage human MSCs at very low densities, were heterogeneous in size, morphology, and potential to differen-tiate. Subcutaneous implantation of single-colony derived strains of human MSCs into immunodeficient mice resulted in the formation of new bone in approximately 58% of the clones, reflecting groups of cells at various levels of differentiation. At least two morphologically distinct cells are present in these cultures: type I cells are spindle shaped and resemble fibroblasts [76]. These cells grow rapidly and represent a subset of the most undifferentiated MSCs and have the greatest potential for self-renewal, culture expansion and differentiation [72]. The type II cells, which are larger and have a broader morphology, divided very slowly. After several passages, the ratio between type I and type II cells changes. The amount of type II cells, which probably arise from the type I cells after asymmetric cell division, in culture increases, and the amount of type I cells, which are the predominant cell type in primary cultures, decreases. In addition, cells with intermediate morphologies are observed. Type I MSCs can be extensively expanded *in vitro* more than 10^4-fold (20–25 passages), before they display any significant changes in morphology, expansion potential or immunophenotype and they can, when cultured under specific conditions, retain their multipotentiality [70]. In addition to the above described type I and II cells, Colter and colleagues described the presence of extremely small cells that were rapidly self-renewing (RS cells) [121] after plating of the MSCs at very low cell densities [123]. Staining with the cell-cycle specific antigen K_i-67 demonstrated that a subset of small, agranular cells (RS-1) was not in cell cycle, whereas the small, granular (RS-2) and the large and moderately granular cells (mature MSCs) were. All three isolated subsets of cells in the MSC cultures were consistently negative for markers of hematopoietic cells, such as CD34 (early HSCs), and CD11b, CD43, CD45 (mature hematopoietic cells). However, some marked differences between epitope expressions of these subsets were observed. In general, less then 10% of all three cell types were dimly positive for the endothelial cell marker CD31 and CD38 (marker for macrophages, B-, T- and NK cells). Mature MSCs were dimly to moderately positive for STRO-1 and CD117 (c-Kit), whereas RS-1 and RS-2 cell populations were negative. Mature MSCs were also positive for CD90 (Thy-1) and expressed the receptors for PDGF and EGF, whereas RS-1 cells were only dimly positive for CD90, and the RS-2 cells completely negative [122, 123]. In contrast, RS cells were positive for VEGF-R2 (Flk-1), TRK, transferrin receptor and annexin II [122]. During the lag phase of growth, RS-1 cells could give rise to RS-2 cells, which in turn gave rise to the more mature MSCs during the log phase, but were also able during the late log phase to regenerate the RS-1 cells. Thus, RS-1 cells, which could be distinguished from more mature MSCs, present in the same cultures, by means of surface marker and protein expression, remained able for a prolonged time in culture to generate single-cell colonies, and showed an enhanced potential for multilineage differentiation [122, 123].

8.2. Phenotype

It is evident that in order to accomplish changes or switches in fate directions of BM-MSCs towards particular differentiation lineages *in vitro*, the initial phenotype of these stem cells must be fully defined. And consequently, along with the elucidation of the phenotype, a better insight in the homing patterns and differentiation programs *in vivo* can be achieved. Although a range of antibodies, raised against multiple cell surface epitopes, can be employed to enrich for MSCs, specific and unique molecular markers are currently not available to unequivocally identify these cells (Table 1.2) [125, 129]. MSCs have been characterized phenotypically in humans as being of non-hematopoietic origin by their lack of expression of hematopoietic markers, such as CD14, CD34 or CD45 [11, 17]. Although the hematopoietic stem cell marker CD34 is not expressed by *ex vivo* culture expanded mesenchymal stem cells, it is possible to directly isolate early MSCs from fresh bone marrow on the basis of CD34 expression [124].

Table 1.2. Main characteristics of human bone marrow-derived MSC*.

Marker type	Antigen	Detection	References
Specific antigens			17, 69, 98, 124
SH2 and SH3	CD73	Pos	
SH4 (endoglin)	CD105	Pos	
STRO-1		Pos	
ASMA	α-smooth muscle actin	Pos	
MAB 1470	Endothelial specific antigen	Pos	
SB-10 (ALCAM)	CD166	Pos	
Cytokine and Growth			11, 17, 30, 31, 128
Factor Receptors			
IL-1Rα, β	CD121a, b	Pos	
IL-2R	CD25	Neg	
IL-3R	CD123	Po	
IL-4R	CD124	Pos	
IL-6R	CD126	Pos	
IL-7R	CD127	Pos	
Transferrin receptor	CD71	Pos	
NGFR		Pos	
TNFα1R	CD120a	Pos	
TNFα2R	CD120b	Pos	
IFNγR	CDw119	Pos	
bFGFR		Pos	
PDGFR	CD140a	Pos	
EGFR		Pos	
LIFR		Pos	
SCFR		Pos	
G-CSFR		Pos	
TGFβ1R		Pos	
TGFβ2R		Pos	
Adhesion molecules (1)			11, 17, 30, 31
Integrins:			
αvβ3	CD51/CD61	Pos	
αvβ5	CD51/b5	Pos	
Integrin subunits:			
VLA-α1	CD49a	Pos	
VLA-α2	CD49b	Pos	
VLA-α3	CD49c	Pos	
VLA-α4	CD49d	Neg	
VLA-α5	CD49e	Pos	
VLA-α6	CD49f	Pos	
VLA-β1	CD29	Pos	
VLA-β3		Pos	
VLA-β4	CD104	Pos	
Adhesion molecules (2)			11, 17, 30, 31
Cadherin 5	CD144	Neg	

(Cont.)

Table 1.2. (Cont.)

Marker type	Antigen	Detection	References
PECAM-1	CD31	Neg	
ICAM-1	CD54	Pos	
ICAM-2	CD102	Pos	
ICAM-3	CD50	Pos	
HCAM	CD44	Pos	
VCAM	CD106	Pos	
NCAM	CD56	Neg	
ALCAM (SB-10)	CD166	Pos	
E-selectin	CD62E	Neg	
L-selectin	CD62L	Pos	
P-selectin	CD62P	Neg	
LFA-1α chain	CD11a	Neg	
LFA-1β chain	CD18	Neg	
LFA-3	CD58	Pos	
CR4 α chain	CD11c	Neg	
Mac 1	CD11b	Neg	
Extracellular matrix			17, 129
Collagen type I, III, IV, V and VI		Pos	
Fibronectin		Pos	
Laminin		Pos	
Hyaluronan		Pos	
Proteoglycan		Pos	
Additional markers			11, 17, 30
T6	CD1a	Neg	
CD3 complex	CD3	Neg	
T4, T8	CD4, CD8	Neg	
Tetraspan	CD9	Neg	
LPS receptor	CD14	Neg	
Lewis X	CD15	Neg	
–	CD34	Neg	
Leukocyte common antigen	CD45	Neg	
B7-1			
HB-15	CD80	Neg	
B7-2	CD83	Neg	
Endothelial specific antigen	CD86	Neg	
Thy-1	CD90	Neg	
MUC18	CD146	Pos	
BST-1	CD157	Pos	
vWF		Pos	
Pan cytokeratins	Pan CK	Pos	
Cytokeratin 18	CK18	Pos	
Cytokeratin 19	CK19	Pos	

*Adapted from reference 30.

However, upon culture expansion, the CD34 antigen is rapidly lost and can only, if at all, be found in very low amounts at the time of the first passage [11, 30, 124]. MSCs are also negative for the dendritic cell marker CD1a, epithelial specific actin (ESA), endothelial cell marker CD31 and CD56 [17]. Human MSCs can be identified by flow cytometry using the monoclonal antibodies SH-2, SH-3 and SH-4, which were originally discovered by Haynesworth et al. [125]. SH-2 recognizes an epitope on endoglin (CD105, receptor for TGF-β1 and TGF-β 3) [126], whereas SH-3 and SH-4 [17, 125] are now known to recognize two distinct epitopes on CD73 (ecto-5'-nucleotidase) [127]. Although neither CD105 (present on T- and B-cells, dendritic cells, endothelial cells, epithelial cells and stem/precursor cells, absent on NK cells, monocytes/macrophages, granulocytes, platelets and erythrocytes) nor CD73 (present on stem/precursor cells, monocytes/macrophages and endothelial cells, absent on T- and B-cells, NK cells, granulocytes, platelets and erythrocytes) are specific for MSCs, the double positive cell populations are highly enriched for MSCs [128]. The antibody SB-10 reacts against activated leukocyte-cell adhesion molecule (ALCAM) and is expressed on early rat, rabbit, canine and human MSCs. During lineage progression, the expression of ALCAM is gradually down regulated and cannot longer be demonstrated on terminally differentiated phenotypes [28, 129]. In addition to ALCAM, MSCs demonstrate a homogenous expression of other adhesion molecules, such as VCAM-1, ICAM-1, -2 and -3, integrins ($\alpha\nu\beta3$ and $\alpha\nu\beta5$) and integrin subunits $\alpha1$, $\alpha2$, $\alpha3$, $\alpha5$, $\alpha6$, $\beta1$, $\beta3$ and $\beta4$, with the exception of subunit $\alpha4$ (CD49d). The expression of several receptors involved in interactions with endothelial cells and ECM components, in particular adhesion molecule CD44 or HCAM (receptor for both hyaluronan and osteopontin) and the strong positive staining for numerous extracellular matrix (ECM) proteins, such as fibronectin, laminin, collagen I, III, IV, V and VI and proteoglycans [16, 17, 130], emphasizes the important key role of MSCs in the regulation of ECM components, concerned with the formation of an optimal niche for hematopoietic cells [31]. Additionally, more than 90% of the human MSCs are recognized by monoclonal antibodies directed against α-smooth muscle actin (ASMA), von Willebrand Factor (vWF), cytokeratins (pan CK, CK18, and CK19) and by MAB 1470, an endothelial specific antigen that exhibits cross reactivity with ASMA-positive cells [17].

Furthermore, the monoclonal antibody STRO-1, discovered by Simmons et al., identifies a distinct population of stromal progenitor cells, present in adult human bone marrow [99, 100] and has been utilized to enrich for CFU-F. However, the surface antigen to which this antibody is directed has not yet been reported. The total amount of STRO-1 positive cells present in the BM represent less than 3 to 5% of the total bone marrow stromal population [97, 99]. The antigen for STRO-1 is expressed by about 10% of the bone marrow MNCs, but most of these cells (about 90%) represent either nucleated erythroid precursor cells (Glycophorin A positive) or a subset of CD19 positive B cells. Interestingly, essentially all CFU-F are present in the STRO-1 positive fraction and MSCs cannot be cultured from the STRO-1 negative subset of cells [99]. The phenotype of the STRO-1 selected CFU-F precursors resembles that of plastic-adherent MSCs and is negative for CD34 and CD45 and positive for CD90, CD106, CD29/CD49, CD10, CD13, and the receptors for PDGF, EGF, insulin-like growth factor (IGF-1) and nerve growth factor (NGF) [30, 77, 131].

MSCs also express type I and II TGF-β receptors [132]. This property was used by Gordon et al. [133] to isolate a stromal precursor from human marrow. Bone marrow cells were cultured under low serum conditions on a collagen matrix, containing a fusion protein, build up out of the receptor-binding domain of TGF-β and a collagen-binding domain, derived from von Willebrand factor (TGF-beta1-vWF). Stromal cells that bound to the engineered peptide were selectively expanded and were found to be able to generate osteogenic colonies after differentiation induction. Similarly, monoclonal antibodies against the low-affinity nerve growth factor receptor (α-LNGFR) can be utilized to label and to obtain a homogenous subset of primitive stromal cells from adult human bone marrow. The NGFR positive fraction was highly enriched for CFU-F compared to the NGFR negative fraction, and showed a 1–3 log larger expansion, in addition to a greater differentiation potential towards osteogenic and adipogenic lineages [134]. LNGFR is highly expressed on freshly isolated and culture expanded BM-MSCs, but is no longer present after terminal differentiation of the cells into fibroblasts, osteoblasts or adipogenic cells. Besides the above-mentioned antibodies, a variety of other antibodies against membrane antigens expressed by MSCs have been utilized to enrich for progenitor cells. These antibodies include CD105 [84], Thy-1 [135], VCAM-1[136], α1-integrin subunit [137], and MUC-18/CD146 [138]. Mesenchymal stem cells express numerous receptors important for cell adhesion with hematopoietic cells (for details see Table 1.2). The production of a broad spectrum of matrix molecules and the extended cytokine expression profile described for MSCs, including production of several hematopoietic and non-hematopoietic growth factors, interleukins, and chemokines (described in the following section), suggests an active and dynamic participation of MSCs in the BM. Here, MSCs are engaged in the formation of a functionally active stromal microenvironment and regulate expansion and differentiation of both hematopoietic and mesenchymal stem cells by the local production of autocrine and paracrine signals [31].

In order to facilitate the selection and standardized use of MSCs, phenotypical description and functional characterization has been the subject of many studies. Although the presence of some typical mesenchymal stem cell markers, for example, SH2, SH3, SH4, HCAM and ALCAM was confirmed by most of these studies, the presence of other markers remains a point of discussion and results are inconclusive. BM-MSCs have been found to be both CD49d negative [11, 30, 51] and dimly positive [17, 139]. As discussed above, human BM-MSCs have been described as both CD34 positive [124] and negative [11, 17]. However, CD34 positive MSCs could only be isolated from fresh BM and never after culture. Therefore, one of the most likely causes of these discrepancies is probably a difference in the proliferative stage [123]. Although MSCs can be expanded extensively and submitted to multiple rounds of passages, this cannot be done without some loss of proliferation capacity and multipotentiality. The remaining cells will start to display changes in morphology and may approach senescence or become apoptotic. Consequently, phenotypical changes will occur, such as the loss of specific cell surface antigens (ALCAM, SH3, SH4, ICAM-1, integrin β1) and result in a functional impairment, as evident from their decreased production of ECM molecules [17]. A second reason may be differences in technical procedures and isolation methods. The already discussed rapidly self-replicating RS cells share most epitopes with MSCs, but additionally express the vascular endothelial growth factor receptor-2 (FLK-1) and annexin II (lipocortin 2), while being negative for STRO-1 [122, 123]. Alternatively, both inter-donor and intra-donor heterogeneity of the obtained BM aspirates may affect the phenotype of the MSCs, just as it affects growth rate [71]. From this point of view it therefore is conceivable that MSCs derived from other tissue sources may show a slightly different expression pattern of surface proteins. In keeping with this, the protein expression of adipose tissue-derived stromal cells is similar to, but not identical to that of reported BM-derived MSCs [51, 140]. Whereas adipose-derived stromal cells are reported as both STRO-1 negative [140] and positive [51], early BM-MSCs are reported as typically positive. Furthermore, a number of studies have reported that AMSCs stain positive for CD49d, whereas the presence of VCAM expression on AMSCs has been found to be both absent [51] and present [140] by different investigators.

In summary, there is still no reliable phenotype to allow prospective isolation of purified MSCs by FACS analysis. Negative depletion of a number of markers (for example, against erythrocytes, endothelial cells, macrophages and monocytes) can be used to enrich for MSCs and to study their properties *in vitro*, but is not sufficient to determine the exact proportion of stem cells and more mature precursor cells in BM, nor is it possible to localize their anatomical site in the bone marrow cavity and other tissues or to track their trafficking. Currently used antibodies are not specific for mesenchymal stem cells, but reflect the presence of growth factors receptors, integrins, adhesion molecules and extracellular matrix components and are therefore also found on a number of other cell types. As a consequence, the isolation of a more specific antibody or antibodies directed against MSCs must remain a priority in ongoing and future research.

8.3. Production of cytokines and growth factors

The stromal layers that develop in primary LTBMC are extraordinarily complex. Analysis of immortalized cell lines obtained from these cultures has shown that the stromal population is heterogeneous and suggested that the stromal cells that maintain long-term repopulating stem cells are rare [141]. Consequently, it seems highly unlikely that all true MSCs are equally involved in maintaining the normal bone and marrow physiology and it is seems more rational to assume that only a relatively small group of cells maintains the steady-state levels in the marrow cavity, whereas the bulk of the remaining cells serve as an emergency backup in case of crisis situations [7]. The cells, responsible for the daily sustenance of cell numbers, represent probably a group of more restricted precursors, which can respond quickly in response to the demands of the bone marrow environment. Therefore, MSCs constitutively secrete specific growth factors and cytokines during each stage of maturation and differentiation into distinct pathways. This involves the rapid modulation of the production of these secreted molecules and the regulation of other signal proteins in a lineage path and stage-specific manner. In addition, MSCs do not only influence their local environment (paracrine effects), they also respond to the cytokines that they themselves produce (autocrine stimulation), implying the presence of complex feedback mechanisms. As a result of this local

Table 1.3. Expression of cytokines by human and mouse MSCs.

Cytokine	Unstimul. hMSC	Unstimul. mMSC	Effect of Dexa on hMSC	Effect of IL-1α on hMSC	Effect of IL-1α or TNF-α on mMSC
IL-1α	−			↑	
IL-2	−			−	
IL-3	−	−		−	
IL-4	−			−	
IL-6	+	+	↓	↑	↑
IL-7	+	+		=	↑
IL-8	+			↑	
IL-10	−			−	
IL-11	+		↓	↑	
IL-12	+			=	
IL-13	−			−	
IL-14	+			=	
IL-15	+	+		=	
LIF	+	+	↓	↑	↑
M-CSF	+		=	↑	↑
G-CSF	+		=	↑	
GM-CSF	−	+		↑	
Flt-3 ligand	+	+		=	
SCF	+	+	=	=	↓
TGF-β1	+	+		↑	↑
TGF-β2	−				
OSM	−	+			
SDF-1	+	+			↓
TPO	+	−			
TNF-α	−	+			
TNF-β	+	+			
IFN-γ	+	−			
IFN-β					

The constitutive mRNA expression of each of the cytokines is marked as + (significant expression) or as − (no basal expression). Effects of Dexamethason, IL-1α and TNF-α on mRNA production are also shown. An upward arrow (↑) indicates increased or newly induced mRNA expression, whereas a downward arrow (↓) indicates decreased mRNA expression. The (=) marks cytokines of which mRNA expression was not markedly altered (10, 28, 78, 144).

signalling, MSCs can induce differentiation in their neighboring cells, giving rise to areas of focal differentiation and the formation of nonuniform colonies [10].

Bone marrow MSCs produce a wide spectrum of growth factors and cytokines, of which most have an important role in the regulation of hematopoiesis [142] (see Table 1.3). Factors detected in primary stromal cell cultures or in stromal cell lines include Granulocyte-Colony Stimulating Factor (G-CSF), GM-CSF, M-CSF,

Interleukin-1 (IL-1), IL-3, IL-6, IL-7, IL-8, IL-11, IL-12, IL-14, IL-15, LIF, fibroblast growth factor (FGF), stem cell factor (SCF), Flt-3 ligand (FL), and tumor necrosis factor (TNF) [10, 28, 143, 144]. Antagonists of hematopoiesis Interferon-γ(IFN-γ), transforming growth factor-β (TGF-β), and MIP-1α, can also be detected in MSC cultures. Haynesworth et al. demonstrated that addition of dexamethasone to the standard growth medium of MSCs decreased the mRNA expression of LIF, IL-6 and IL-11 [144]. In contrast, in response to IL-1α the expression of G-CSF, M-CSF, LIF, IL-6 and IL-11 increased and expression of GM-CSF was induced [28, 143, 144]. Thalmeier et al, who analysed the cytokine expression patterns of two permanent human bone marrow stromal cell lines, L87/4 and L88/5 [145], obtained similar results. Constitutive mRNA expression of c-kit, G-CSF, GM-CSF, IL-1β, IL-6, IL-7, IL-8, IL-11, SCF, LIF, M-CSF, MIP-1α, TGF-β, and TNF-α was demonstrated in both cell lines. Irradiation and IL-1α treatment induced an increase in mRNA levels for GM-CSF, IL-1β, and LIF by in both cell lines, as confirmed northern blot analysis. Effects induced by IL-1α treatment on GM-CSF, IL-1 beta, IL-6, IL-11, and LIF mRNA levels could be antagonized by addition of dexamethasone. In contrast, dexamethasone did not affect the levels of IL-1α-induced G-CSF mRNA. Both cell lines showed an increase in SCF mRNA, after stimulation with dexamethasone, but not in response to IL-1α.

8.4. Cell cycle status

Information about the cell cycle status of MSCs can be obtained by permeabilization of the cells, measurement of the DNA content by labeling with Propidium Iodide (PI) and subsequent analysis by flow cytometry. PI staining revealed that most cells (more than 90%) were either in G0 (growth arrested) or G1 phase (actively growing), whereas only a small population of cells (10%) was actively proliferating (S+G2+M). Labeling of the DNA and RNA with acridine orange demonstrated that approximately 20% of MSCs were actually quiescent, non-dividing cells [17].

9. MOBILIZATION AND MICROENVIRONMENT OF MSCS

It is crucial to learn whether the MSCs detected in various mesenchymal tissues are inherent there, or whether their pool is replenished, and if so to what extent, by constant or on demand migration of mesenchymal stem and progenitor cells from the BM. It was recently shown that MSCs are not only present in adult BM and other mesenchymal tissues, but also in umbilical cord blood. Similar to adult BM-MSCs, UCB-derived MSCs can be obtained by their characteristic ability to adhere to plastic surfaces. Furthermore, they also share some of the other specific properties of adult BM-MSCs, including similarities in morphology, immunophenotype, and differentiation potential [22]. The presence of MSCs in UCB was shown to be inversely proportional to fetal age, suggesting that MSCs somehow migrate out of the cord into the different tissues during early fetal ontogeny (first trimester) [36]. As a consequence, no MSCs could be cultured from UCB at later stages of development. Apart from distant migration patterns along blood vessels, MSCs can migrate locally into tissues in response to signals induced by local injury and participate in support and repair of tissues, such as cartilage and (heart) muscle [90]. Primitive MSCs, present in the BM, can serve as a regular supplier of committed mesenchymal cells to distant tissues [16, 68]. In addition, committed mesenchymal precursors with restricted differentiation potentials and uncommitted MSCs are also intrinsically present in marrow-distant mesenchymal tissues, such as in muscle [56, 59], fatty tissue [47, 49, 51] and bone [52, 53, 54].

To be able to serve as stem cells for distant tissues, BM-MSCs must be able to leave the marrow microenvironment after either symmetric division (self-renewal), after which the one stem daughter remains, and the second travels, or after asymmetric division, giving rise to a lineage-committed daughter cell that travels. Lineage commitment does not necessarily occur at one particular site, but in stead appears to be the result of multiple cell-cell interactions and local signals, derived from the stem cell niche, through which the cells travels on its way to its final destination [9, 19]. This niche or stem cell microenvironment is formed by a subset of tissue cells combined with the external signals, where they direct the stem cell destiny conjointly. The niche serves as an important meeting place for the exchange of information between uncommitted precursors, their progeny and their adjacent cells [23]. In addition to interactions between mesenchymal and non-mesenchymal cells, their excretion products (growth factors and ECM molecules) and other specific modulators of differentiation take part in the construction of spatial and temporal relationships in the MSC niche. As a consequence, tight cell-to-cell or cell-to-matrix attachments between mesenchymal progenitors and the surrounding stromal elements should relax in order to promote the release of the cells into the peripheral circulation. Here, the progenitor cells find themselves in a temporary compartment, which serves as a medium to guide the cells to their new environment, where they can participate in local processes [31]. This hypothesis seems feasible, but how should it then be explained that no convincing evidence has been offered to demonstrate the presence of circulating MSCs in peripheral blood collections beyond any doubt? Several groups have attempted to culture plastic-adherent progenitor cells from PB with varying success [21, 40, 41, 43], as discussed above. From these combined data, it becomes increasingly clear that if MSCs circulate in the PB, they do so in very low numbers and most probably only after stimulation with cytokines from either an *ex vivo* source (for example, after G-CSF or GM-CSF stimulation) or from an *in vivo* source, where a gradient or a combination of cytokines leads the MSC to its new niche (usually a place of significantly severe injury). It therefore appears relevant to firstly further determine whether mesenchymal stem and/or progenitor cells normally circulate in the blood of healthy individuals or during specific disease states and, secondly, if a significant number of these cells can be actively induced to enter the blood stream by means of cytokine stimulation. This is commonly known to occur with HSCs, which are primed from the BM into the PB after treatment with cytotoxic agents or growth factors [24].

The maintenance of stem cell compartments is in general regulated by cell autonomous regulators, which in turn are modulated by external signals [19]. These intrinsic regulators are comprised of factors involved in asymmetrical cell division, the expression of differentiation-related genes and signals determining cell division and telomerase activity [19, 31]. However, the specific characteristics and properties of the microenvironment for MSCs are just starting to be unraveled and are not yet known in detail. In the long term, self-renewal may become less important for the maintenance of a physiological state in mesenchymal tissues than the capacity to differentiate into a multiplicity of lineages or phenotypic flexibility if commitment and differentiation are actually reversible *in vivo* in response to environmental signals [3, 192], as has now been observed to be the case under special culture conditions *in vitro* [109, 113]. These data suggest that even if BM-MSCs can be induced to circulate and leave their niche, this is not necessarily in contradiction with the presence of inherent mesenchymal stem cells residing in the various mesenchymal tissues. Moreover, there appears to be some state of equilibrium, in which both kinds of stem cells (resident and distant) are co-operating and the resident stem cells maintain the daily requests and demands of the tissue, whereas in case of emergency and significant tissue injury, MSCs from distant tissues such as the BM can be recruited.

10. GENETIC ENGINEERING OF MSCS

MSCs are of great therapeutic potential because they are easily isolated from a small aspirate and readily generate single-cell-derived colonies. In addition, these cells are of special interest to their ability to self-renew and differentiate into multiple tissues. MSCs are relatively easy to expand in culture and can be transduced with exogenous genes without the need for addition of cytokines and therefore appear to have several advantages over the use of HSCs in gene therapy. Additionally, transduction of MSCs does not seem to impair their ability to home to several hematopoietic organs (i.e. BM, spleen, liver) [146, 147], nor does it appear to affect their ability for self-renewal [147]. For these reasons, the cells are a potential powerful device for tissue engineering and are currently being tested for their potential use in cell and gene therapy for a number of different kinds of diseases. The following step towards *in vivo* use of genetically engineered MSCs is the development and use of animal transplantation models for bone, cartilage, tendon, marrow stroma, and muscle repair and regeneration [148].

In a number of studies of gene transfer into MSCs, it was demonstrated that both animal and human adherent stromal cells could be efficiently transduced with exogenous genes employing a variety of distinct vectors, without an evident effect on their stem cell characteristics, as shown by maintained ability for lineage progression into several phenotypes and by transplantation studies, in which long-term expression of the transgenes was observed [149, 150, 151, 152]. Furthermore, it was demonstrated that the transfer of genes into human MSCs by use of retroviral vectors resulted in long-term *in vitro* and *in vivo* expression of GFP [153, 154, 155]. Others obtained similar results after introduction of the genes for LacZ and neoR [147, 149, 152, 156, 157]. MSCs, transduced with genes for growth factors and cytokines, were shown to express the proteins both *in vitro* and *in vivo* [150, 151, 153, 158, 159, 160] and to possess a couple of thousand copies of transgene mRNA per cell, detectable for up to 6 months after infusion [151]. Subsequently, transduction protocols have been developed to reach almost homogenous transduction percentages of about 80 to 90% [151]. By applying retroviral transduction to animal-derived MSCs, Mosca et al. were able to successfully transduce MSCs obtained from seven non-human species, including baboon, canine, rat, sheep, goat, pig and rabbit [151]. Although MSCs from all species could be transduced, the absence of specific amphotropic retroviral receptors on MSCs derived from pig, sheep, goat and rabbit MSCs resulted in little or no transgene expression, whereas use of the same retrovirus resulted in a transduction efficiency of more than 80% for human MSCs. The feasibility of adenovirus-mediated (Adv) gene transfer into human MSCs has been assessed by Conget et al. [161] They found that after transduction with replication-defective Adv-containing reporter genes (LacZ or GFP) under the control of a CMV promoter, only a subset of cells (approximately 20%) expressed the transgenes at high levels. Although infection was only observed in cells expressing both Adv-attachment (CAR) and Adv-internalization (integrin $\alpha\nu\beta5$) receptors, gene transfer efficiency was determined mainly by the limited levels of expression of CAR. Importantly, the differentiation potential of the transduced cells was changed as no adipogenic differentiation could be achieved, whereas osteogenic potential was maintained.

Several different strategies are under investigation for the direct therapeutic use of MSCs. For example, MSCs can be isolated from patients with degenerative osteoarthritis, expanded and utilized for resurfacing of the affected joints. Alternatively, MSCs can be transplanted into poorly healing bone fractures, cartilage and tendon defects, but also into damaged heart muscle, where they can either actively participate in or merely support the repair processes [162, 163, 164]. Indirectly, by transducing MSCs with exogenous genes, the genetically marked MSCs can be followed *in vivo*, and normal functioning genes can be transferred into cells with dysfunctional mutations and supply the lacking protein product. For instance, gene therapy could be employed for the correction of osteogenesis imperfecta (brittle bone disease) by transplantation of (autologous) MSCs containing a wild-type gene for type I collagen [83, 165, 166]. Long term cultured bone marrow stromal cells (more than 20 passages) obtained from a mouse model for osteogenesis imperfecta (OI) and retrovirally transduced with LacZ and neoR genes were shown to form bone and to express exogenous genes after direct infusion into femurs, but were also capable of trafficking through the circulatory system and home to the contra lateral bones [156]. Saito et al. injected BM-MSCs of healthy, congenic mouse into mdx mice, which have a point mutation in the gene for dystrophin are being used as a mouse model for its human counterpart Duchenne's muscular dystrophy. The injected MSCs differentiated into functional muscle cells and fused with host myotubes, resulting in the expression of dystrophin [167]. Duchenne's disease therefore, could possibly be treated effectively by transduction of autologous MSCs and inserting the normal gene. Keating et al. reported that human MSCs, transfected by electroporation with hCMVhFIXcDNA, a construct containing the gene for factor IX, secrete the protein product for up to 12 weeks after infusion into SCID mice [168]. Transfection of expanded canine MSCs with a human factor IX (hFIX) plasmid vector resulted in detectable hFIX levels in plasma for at least 9 days [169]. In addition, transplantation of human MSCs, engineered to express factor VIII, into NOD-SCID mice resulted in expression of the protein up to 3 weeks post injection [170]. Although the human fXIII plasma expression disappeared, most probably due to promoter inactivation, a small group of transduced cells remained detectable for at least 4 months. Therefore, it seems feasible to make use of engineered MSCs as a delivery vehicle of genes for the treatment of both hemophilia A and B and possibly other genetic diseases caused by deficiencies in circulating proteins. In an attempt to obtain enhanced human hematopoiesis in bnx mice, Dao and colleagues co-transplanted CD34+ cells with MSCs engineered to secrete several human cytokines, including IL-3, IL-7, Epo, FLT and SCF [160]. Sustained secretion of human cytokines into the murine bloodstream at supraphysiological levels was detected for up to 6 months after injection, but detectable human hematopoiesis was supported only in the presence of IL-3 production. It seems feasible to combine both direct and indirect applications of MSCs by engineering these cells to enhance the effect of their own transplantation by local expression of desired therapeutic proteins. For example, autologous MSCs, obtained from patients suffering from osteoporosis can be transduced with a gene for BMP-2 that after transplantation induces an autocrine regulated differentiation into an osteogenic pathway [171].

Recently a study was published by Campagnoli and colleagues, who reported that MSCs could be derived from fetal tissues, such as blood, liver and bone marrow, from as early as eight weeks of gestation. They demonstrated that these cells could be relatively easy retrovirally transduced, with more than 99% of the cells expressing enhanced GFP (EGFP) at high levels, without any kind of selection [172]. They suggested that MSCs could therefore also serve as suitable vehicles for the prenatal delivery of certain genes and even more boldly, that genetically modified autologous fetal MSCs could be possibly used as an alternative treatment modality for genetic disorders, known to cause irreversible damage before birth, when transplanted *in utero*.

The previously described reports thus provide convincing evidence that mesenchymal stem and progenitor cell populations, derived from bone marrow and other tissues, from both prenatal and postnatal sources, and obtained from a number of different species, can be genetically engineered to express a spectrum of different proteins *in vitro* and *in vivo*, at significant levels for extended periods of time. These genetically modified MSCs can be potentially utilized to treat a variety of genetic or acquired diseases, including protein deficiencies, bone, cartilage, (cardio-) myogenic and BM stromal disorders, neurological diseases, such as Parkinson's disease, multiple sclerosis, or cerebro-vascular accidents and perhaps even malignancies [173].

Although high transduction efficiencies have been obtained by the current protocols, further research should focus on promoting the stable long-term expression of these genes and prevention of silencing and/or decrease of protein expression. Prolonged *ex vivo* expansion (3–4 weeks) to increase cell numbers was found to reduce the transduction efficiency in MSCs, whereas MSCs maintained in culture for only 10–12 days were successfully transduced [146, 157]. However, in order to obtain sufficiently high transduction percentages, up to four rounds of incubation with virus-containing supernatant may prove to be necessary [146, 149]. Most studies undertaken thus far reported to maintain various levels of expression of the inserted gene for periods ranging from several days up to 6 months [147, 155, 160]. However, expression of transgenes in mouse MSCs extinguished with time in most studies, most probably due to promoter inactivation as a result of DNA methylation [170]. In contrast, the diminished expression of transgenes in human MSCs resulted most likely from a loss of transduced cells [46].

If only a temporary or slowly extinguishing protein expression is intended, for example, to support local bone repair in case of fractures, a less permanent solution can be achieved by means of electroporation [168], calcium phosphate precipitation, lipofection, or use of plasmids and Adv constructs [46, 161]. Adenoviral-mediated infections have the advantage over the use of retroviruses since they do not require cell divisions for gene insertion and have a low toxicity. As a corollary, multiple copies of the intended gene can be inserted into the genome of the target cell. However, the number of copies transferred to the transfected MSCs cannot be controlled and is therefore highly variable and unpredictable. Another major issue to address is the how to target the genetically engineered cells to the desired tissues, which includes the maintenance of their ability to home and functionally engraft in significant numbers to be of any temporary (tissue repair) or sustained (protein replacement) clinical efficacy.

11. CLINICAL USE AND TRANSPLANTATION OF MSC

Recipients of an unmanipulated allogeneic bone marrow transplant contain only host-type marrow stromal cells and MSCs in their bone marrow [73, 174, 175, 176, 177]. The characteristic properties and viability of stromal cells are not affected when stromal layers are *in vitro* subjected to a low dose of irradiation [174]. The absence of donor-derived MSCs can therefore be ascribed at least partially, to the inability of the conditioning regimen to ablate the host marrow stroma and create a place for the donor cells to grow. In addition, the number of transplanted MSCs in an average bone marrow graft is estimated to be approximately 2 to 5 MSCs per 1×10^6 MNCs [73] and may therefore not be sufficient to obtain any substantial engraftment. Although an adherent cell layer containing a subpopulation (2–44%) of donor-derived cells could be obtained from the BM of patients with 100% donor hematopoiesis after an allogeneic HSC infusion, these cells were positive for CD14 and CD45 indicating a macrophage phenotype and expressed non-specific esterase [178]. However, the function of the recipient stromal cells is often poor due to disruption of the stem cell niches in the marrow cavity by hemorrhaging and loss of fat deposits and connective tissue elements as a result of the intensive therapy regimen consisting of cytotoxic radio- and/or chemotherapy [179, 180, 182]. Once the stromal environment has been seriously damaged, the function of the stromal cells remains persistently impaired over a long time [179]. This damaged stroma may not be sufficiently capable of supporting of growth of hematopoietic stem and progenitor cells after infusion and therefore, reconstitution of the bone marrow stromal compartment may enhance and facilitate the restoration of hematopoiesis [180]. Koç et al. intravenously injected up to 2.2×10^6 human culture-expanded autologous MSCs per kg within 1 to 24 hours after infusion of PB-derived progenitor cells into advanced breast cancer patients after high dose chemotherapy and found a rapid hematopoietic recovery, without any signs of infusion-related toxicity or other side effects [181]. Although they suggested that transplantation of MSCs in combination with HSCs enhanced the engraftment of the HSCs, the daily administration of G-CSF to the patients until neutrophil engraftment, must undoubtedly have had some stimulating effect too, and should not be neglected. In mice, stromal chimerism was achieved after infusions of genetically marked culture-expanded stromal progenitors. Donor MSCs obtained from transgenic mice expressing a mini-gene for collagen I, were intravenously injected into irradiated FVB/N mice and replicated *in vivo*, populating several connective tissues over a period of weeks to months, including the bone, carti-lage and lung [82]. When healthy FVB/N mice served as MSC donors for transgenic mice with an osteogenesis imperfecta phenotype, the donor-derived stromal cells could be recovered for over 2.5 months from a variety of mesenchymal tissues, such as the lung, calvaria, long bone, cartilage, skin and tail [165]. These encouraging results have stimulated the clinical use of allogeneic MSCs after standard HSC transplantation for the treatment of bone defects in six children with severe forms of osteogenesis imperfecta [83].

The *in vivo* distribution of rat MSCs after intravenous, intra-arterial or intraperi-toneal injection was measured by radioactive labeling of the cells with ^{111}In-oxine [182]. Labeled MSCs were found in liver, lungs, kidneys, spleen and long bones. How-ever, MSCs lodged predominantly to the lung and secondarily to the liver, whereas only a small fraction of the infused cells homed to the BM. After administration of sodium nitroprusside, more ^{111}In-oxine-MSCs passed through the lungs, resulting in a significant increase in homing towards BM and long bones. These results probably reflect one of the major problems of MSC transplantation biology: stromal cells are relatively large (approximately 2 to 3 times the size of a neutrophil [181]) and may not manoeuvre as easy through the circulation as blood cells do. Therefore, administration of a vasodilator can influence the distribution of MSCs and prevent the cells from lodging in the lung capillaries.

Azizi et al. detected the presence of approximately 20% of human marrow-derived MSCs, which were directly injected into the corpus striatum of rat brains, in several layers of the brain without any signs of rejection or inflammation for up to 72 days after infusion. The injected cells showed a migration pattern similar to astrocyte grafts. These data suggested that the brain may also be a suitable target for *in vivo* use of (genetically engineered) MSCs [87]. Culture-expanded MSCs have been tested in pre-clinical models for the repair of bone, cartilage and tendon/ligament by local delivery of MSCs within an appropriate matrix [180]. In a non-human primate model, it was shown that the intravenous infusion of $3–30 \times 10^6$ cells/kg of either unmodified or retrovirally transduced baboon MSCs, into lethally irradiated baboons was not associ-ated with any significant toxicity [183]. The transplanted baboon MSCs were capable of homing to the BM and were able to persist within the BM for up to 1 year after infusion.

In the above discussed series of preclinical transplantation studies in animals and in at least two clinical trials performed, no evidence of systemic infusion-related toxi-city or other adverse reactions were observed after transplantation of *ex vivo* expanded autologous or allogeneic MSCs and therefore MSC transplantation appears feasible and safe at least in the short-term [180, 181]. Cocultures, in which MSCs were used to support expansion and maintenance of HSCs, suggested that unrelated donor MSCs may not generate alloreactive lymphocytes in *in vitro* culture conditions and may have a modulating effect on the immune system *in vivo* [184]. Since MSCs are precursor cells for a variety of mesenchymal lineages and can also be detected in multiple tissues involved in autoimmune diseases, they may have a potential function as target cells for gene-engineered immunomodulation [187]. Baboon and human MSCs failed to elicit a proliferative response from allogeneic lymphocytes and were able to induce a 50%

and 65% reduction, respectively, in proliferative activity when added into a mixed lymphocyte reaction (MLR) or to mitogen-stimulated lymphocytes [184, 185]. In humans however, there appears to be a dose-depending effect, since addition of 10.000–40.000 MSCs to a MLR resulted in suppression of proliferative activity, whereas the addition of smaller amounts of MSCs (10–10.000) led to a less consistent suppression or could even result in stimulation of lymphocyte proliferation [185, 186]. In addition, these responses appeared to be independent of the major histocompatibility complex (MHC), since the addition of "third" party MSCs or MSCs, autologous to the responder or stimulating lymphocytes, to the MLR resulted in a similar dose-dependent inhibition [185, 186]. T-cell inhibition of MSCs due to apoptosis-related mechanisms was excluded. Both T cells and MSCs displayed more than 95% viability as demonstrated by Trypan Blue staining, and T cells were able to resume proliferation once provided with humoral of cellular stimuli [186]. MSC-mediated inhibition was likely due to interactions with soluble factors, since physical separation of MSCs and effector cells by culturing in a transwell system resulted in a similar suppression [185]. Engineered MSCs could therefore possibly serve as a vehicle to distribute locally active therapeutic proteins and to deliver tissue-specific immunosuppressive cytokines in order to improve the efficiency of immunotherapy [187]. MSCs themselves are thought to be able to escape from recognition by the immune system because they lack expression of MHC class II antigens, the T-cell co-stimulatory molecules B7-1 and B7-2, CD40 and CD40L [30]. Consequently, transplantation of MSCs may have a suppressing effect on graft-versus-host disease. Preliminary results indicate that allogeneic MSCs are well tolerated when infused into patients, that they may play indeed an important role in decreasing graft-versus-host reactions and that transplantation of MSCs may have a significant clinical benefit [187, 188, 189]. Recently, a clinical trial was undertaken to assess the feasibility and possible toxicity of transduced allogeneic MSC infusion into children with osteogenesis imperfecta [83]. MSCs were obtained from the siblings of the patients or unrelated donors and retrovirally transduced with either G1PLII or the LNc8 vector after the first passage in culture. The transduction efficiency ranged from 2 to 25% for both vectors in all samples. All patients received a first dose of minimally cultured MSCs (10^6 cells/kg body weight) and a second dose of MSCs, that were expanded over three passages (5×10^6 cells/kg). Five out of six patients showed engraftment of G1PL11 marked MSCs in bone, stroma and/or skin. However, none of the biopsies contained evidence of the LNc8 vector, which encodes for the neo^R gene. Failure to detect LNc8 proviral sequences after transplantation suggested an immunological response elicited against MSCs expressing the Neo^R gene. This was confirmed by the demonstration of a dose-dependent cytotoxic T cell-mediated lysis of LNc8 transduced cells in vitro. No significant lysis was observed in G1PLII or mock-transduced cells. One patient developed an urticarial rash, which rapidly resolved without sequelae. No other clinically significant infusion-related toxicity occurred in any of the other patients. All patients except one displayed a spectacular increase in growth within 6 months after transplantation, ranging from 60% up to 94% of the predicted median. These data suggest that donor MSCs can be safely administered, but the genetically engineered MSCs expressing foreign proteins, such as the Neo^R, may result in an increased risk for immune attack. This implies that if MSCs will be employed as vehicles to deliver therapeutic proteins, these proteins must be recognized by the patient's immune system as non-foreign [83].

In conclusion, potential clinical applications of MSCs may involve (a combination of) the following: replacement and/or enhancement of damaged stroma and stromal functions (including enhancement of hematopoietic recovery after HSC transplantation); tissue specific production and regulation of locally active cytokines, growth factors or other therapeutic proteins (intrinsic expression or induced by genetic engineering); and control or at least modulation of graft versus host disease.

12. SUMMARY

Recently, considerable improvement and gain of knowledge has been obtained with respect to MSC biology, development and clinical options. New methods for selective isolation, *ex vivo* expansion, and assays for evaluation of differentiation potential were developed. Although multiple molecular markers and a variety of antibodies directed against cell surface proteins have become (commercially) available for detection and selection of stromal progenitors and their differentiating progeny, the currently available methods for *in vitro* characterization and *in vivo* follow-up of the real mesenchymal stem cell are rather poor and the markers employed are neither sufficiently specific nor reliable. Consequently, the search for new markers in order to facilitate and standardize the identification of the mesenchymal stem cell continues. Although a number of different methods have been developed to enrich for early stem cells on the basis of phenotypical (surface markers), morphological (size or granularity), or behavioural characteristics (plastic-adherence), the impact of the isolation and/or subsequent cultivation is currently unknown. Efforts undertaken to isolate and to describe the qualities and peculiarities, in addition to the distinctive traits of the "true" mesenchymal stem cell, are few in number. Even more, it has been very difficult to gain an improved insight in the functional characteristics of MSCs, since the tests (for example, differentiation assays) themselves may influence (the properties of) the stem cells and it has been virtually impossible to create an *in vitro* environment, which is an exact representation of the in situ situation, for studying the MSCs.

Clinical use of MSCs for the treatment of a number of genetic, or acquired diseases seems now feasible and MSC transplantation in animal models has provided clues for the incredible potential of the use of these cells. MSCs as a target for gene therapy have several advantages over the use of HSCs. First, unlike HSCs, MSCs are relatively easy to expand without the addition of specific growth factors. Second, MSCs do not have to be prestimulated for transduction to induce the cells to enter the cell cycle. Third, MSCs may functionally engraft in a number of mesenchymal tissues where HSCs cannot reach and may have locally active or supportive effects, or both. Gene engineering of MSCs may have advantages over engineering of HSCs, but as is the case with HSCs, random insertion of genes, implies the possibility of mutagenesis and safety of the procedure must be thoroughly assessed before clinical trials are started.

The presented reports thus seem very promising, but there are several problems that must be attended to. Transplantation kinetics and organ distribution after systemic infusion of MSCs are far from clear and unlike HSCs, have not yet been studied extensively.

Expansion of MSCs *in vitro* prior to transplantation provides the opportunity to infuse higher numbers of MSCs. However, it must be kept in mind that progenitor potential and function may be affected during long-term cultures [17, 72], and therefore, expansion by the stem cells does not necessarily imply expansion of the stem cell group itself. In particular, current research should therefore focus on the identification of the "true" stem cell phenotype permitting the isolation and purification of MSCs by flow cytometry without the additional need for culture expansion.

Feasibility and short-term safety of infusion of limited numbers of MSCs have been demonstrated in a number of studies, but efficacy is still in need of improvement and long-term influences of MSCs infusions must be assessed by intensive preclinical evaluation and carefully monitored follow-up.

In conclusion, the use of MSCs seems full of promises and challenges, and offers the potential for a wide range of new therapeutic options, but fundamental factors are still unclear and remain to be revealed in the forthcoming decades.

ACKNOWLEDGMENT

This work has been in part supported by Contracts of the Commission of the European Communities, in particular G5RD-CT-2002-00738.

REFERENCES

[1] Junquiera LC, Carneiro J, Kelley RO. *Basic Histology.* 7th ed., 1992.
[2] Weiss L, Geduldig U. Barrier cells: stromal regulation of hematopoiesis and blood cell release in normal and stressed murine bone marrow. *Blood.* 1991;78:975–990.
[3] Bianco P, Gehron Robey P. Marrow stromal stem cells. *J Clin Invest.* 2000;105(12):1663–1668.
[4] Bianco P, Riminucci M, Kuznetsov S, Gehron Robey P. Multipotential cells in the bone marrow stroma: regulation in the context of organ physiology. *Crit Rev Eukaryot Gene Expr.* 1999;9:159–173.
[5] Ducy P, Zhang R, Geoffroy V, Ridall AL, Karsenty G. Osf2/Cbfa1: a transcriptional activator of osteoblast differentiation. *Cell.* 1997;89:747–754.
[6] Satomura K, Krebsbach P, Bianco P, Gehron Robey P. Osteogenic imprinting upstream of marrow stromal cell differentiation. *J Cell Biochem.* 2000;78(3):391–403.
[7] Owen M. The marrow stromal cell system. In: Beresford JN, Owen ME, eds. *Marrow stromal Cell Culture.* Cambridge University Press; 1998:chap 1.
[8] Bianco P, Riminucci M. The bone marrow stroma *in vivo*: ontogeny, structure, cellular composition and changes in disease. In: Beresford JN, Owen ME, eds. *Marrow Stromal Cell Culture.* United Kingdom: Cambridge University Press; 1998:chap 2.
[9] Muller-Sieburg CE, Deryugina E. The stromal cells' guide to the stem cell universe. *Stem Cells.* 1995;13:477–486.
[10] Dormady SP, Bashayan O, Dougherty R, Zhang XM, Basch RS. Immortalized multipotential mesenchymal cells and the hematopoietic microenvironment. *J Hematother Stem Cell Res.* 2001;10:125–140.
[11] Pittenger MF, Mackay AM, Beck SC, et al. Multilineage potential of adult human mesenchymal stem cells. *Science.* 1999;284:143–147.
[12] Friedenstein AJ, Chailakhyan RK, Lalykina KS. The development of fibroblast colonies in monolayer cultures of guinea pig bone marrow and spleen cells. *Cell Tissue Kinet.* 1970;3(4):393–403.
[13] Piersma AH, Brockbank KG, Ploemacher RE, van Vliet E, Brakel-van Peer KM, Visser PJ. Characterization of fibroblastic stromal cells from murine bone marrow. *Exp Hematol.* 1985;13:237–243.
[14] Kuznetsov SA, Krebsbach PH, Satomura K, et al. Single-colony derived strains of human marrow stromal fibroblasts form bone after transplantation *in vivo*. *J Bone Miner Res.* 1997;12(9):1335–1347.

[15] Kuznetsov SA, Friedenstein AJ, Gehron Robey P. Factors required for bone marrow stromal fibroblast colony formation *in vitro*. *Br J Haematol*. 1997;97:561–570.
[16] Prockop DJ. Marrow stromal cells as stem cells for nonhematopoietic tissues. *Science*. 1997;276:71–74.
[17] Conget PA, Minguell JJ. Phenotypical and functional properties of human bone marrow mesenchymal progenitor cells. *J Cell Physiol*. 1999;181:67–73.
[18] Fuchs E, Segre J. Stem cells: a new lease on life. *Cell*. 2000;100:143–155.
[19] Watt FM, Hogan BLM. Out of Eden: stem cells and their niches. *Science*. 2000;287:1427–1430.
[20] Tavassoli M, Crosby WH. Transplantation of marrow to extramedullary sites. *Science*. 1968;161(836):54–6.
[21] Fernandez M, Simon V, Herrera G, Cao C, Del Favero H, Minguell JJ. Detection of stromal cells in peripheral blood progenitor cell collections from breast cancer patients. *Bone Marrow Transplant*. 1997;20:265–271.
[22] Erices A, Conget P, Minguell JJ. Mesenchymal progenitor cells in human umbilical cord blood. *Br J Haematol*. 2000;109:235–242.
[23] Sukhikh GT, Malaitsev VV, Bogdanova IM, Dubrovina IV. Mesenchymal stem cells. *Bull Exp Biol Med*. 2002;133(2):103–109.
[24] Shimizu N, Asai T, Hashimoto S, et al. Mobilization factors of peripheral blood stem cells in healthy donors. *Ther Apher*. 2002;6(6):413–418.
[25] Menendez P, Caballero MD, Prosper F, et al. The composition of leukapheresis products impacts on the hematopoietic recovery after autologous transplantation independently of the mobilization regimen. *Transfusion*. 2002;42(9):1159–1172.
[26] Dexter TM. Stromal cell associated haemopoiesis. *J Cell Physiol Suppl*. 1982;1:87–94.
[27] Gartner S, Kaplan HS. Long-term culture of bone marrow cells. *Proc Natl Acad Sci USA*. 1980;77:4756–4759.
[28] Majumdar MK, Thiede MA, Mosca JD, Moorman M, Gerson SL. Phenotypic and functional comparison of cultures of marrow-derived mesenchymal stem cells (MSCs) and stromal cells. *J Cell Physiol*. 1998;176:57–66.
[29] Lodie TA, Blickarz CE, Devarakonda TJ, et al. Systematic analysis of reportedly distinct populations of multipotent bone marrow-derived stem cells reveals a lack of distinction. *Tissue Eng*. 2002;8(5):739–751.
[30] Deans RJ, Moseley AB. Mesenchymal stem cells: biology and potential clinical uses. *Exp Hematol*. 2000;28:875–884.
[31] Minguell JJ, Erices A, Conget P. Mesenchymal stem cells. *Exp Biol Med*. 2001;226(6):507–520.
[32] Mareschi K, Biasin E, Piacibello W, Aglietta M, Madon E, Fagioli F. Isolation of human mesenchymal stem cells: bone marrow versus umbilical cord blood. *Haematologica*. 2001;86(10):1099–1100.
[33] Wexler SA, Donaldson C, Denning-Kendall P, Rice C, Bradley B, Hows JM. Adult bone marrow is a rich source of human mesenchymal "stem"cells but umbilical cord and mobilized adult blood are not. *Br J Haematol*. 2003;121:368–374.
[34] Ye ZQ, Burkholder JK, Qiu P, Schultz JC, Shahidi NT, Yang NS. Establishment of an adherent cell feeder layer from human umbilical cord blood for support of long-term hematopoietic progenitor cell growth. *Proc Natl Acad Sci USA*. 1994;91:12140–12144.
[35] Gutierrez-Rodriguez M, Reyes-Maldonado E, Mayani H. Characterization of the adherent cells developed in Dexter-type long-term cultures from human umbilical cord blood. *Stem Cells*. 2000;18:46–52.
[36] Campagnoli C, Roberts IAG, Kumar S, Bennett PR, Bellantuono I, Fisk NM. Identification of mesenchymal stem/progenitor cells in human first-trimester fetal blood, liver, and bone marrow. *Blood*. 2001;98:2396–2402.
[37] Suda T, Takahashi N, Martin J. Modulation of osteoclast differentiation. *Endocr Rev*. 1992;13:66–80.
[38] Minguell JJ, Conget P, Erices A. Biology and clinical utilization of mesenchymal progenitor cells. *Braz J Med Biol Res*. 2000;33:881–887.
[39] Romanov YA, Svintsitskaya VA, Smirnov VN. Searching for alternative sources of postnatal human mesenchymal stem cells: candidate MSC-like cells from umbilical cord blood. *Stem Cells*. 2003;21:105–110.
[40] Lazarus HM, Haynesworth SE, Gerson SL, Caplan AI. Human bone marrow-derived mesenchymal (stromal) progenitor cells (MPCs) cannot be recovered from peripheral blood progenitor cell collections. *J Hematother*. 1997;6(5):447–455.
[41] Ojeda-Uribe M, Brunot A, Lenat A, Legros M. Failure to detect spindle-shaped fibroblastoid cell progenitors in PBPC collections. *Acta Haematol*. 1993;90(3):139–143.
[42] Purton LE, Mielcarek M, Torok-Storb B. Monocytes are the likely candidate "stromal" cell in G-CSF mobilized peripheral blood. *Bone Marrow Transplant*. 1998;21:1075–1076.

[43] Zvaifler NJ, Marinova-Mutafchieva L, Adams G, et al. Mesenchymal precursors in the blood of normal individuals. *Arthritis Res.* 2000;2:477–488.

[44] Piersma AH, Ploemacher RE, Brockbank KG, Nikkels PG, Ottenheim CP. Migration of fibroblastoid stromal cells in murine blood. *Cell Tissue Kinet.* 1985;18(6):589–595.

[45] Kuznetsov SA, Mankani MH, Gronthos S, Satomura K, Bianco P, Gehron Robey P. Circulating skeletal stem cells. *J Cell Biol.* 2001;153(5):1133–1139.

[46] Bianco P, Riminucci M, Gronthos S, Gehron Robey P. Bone marrow stromal stem cells: nature, biology and potential applications. *Stem Cells.* 2001;19:180–192.

[47] Zuk PA, Zhu M, Mizuno H, et al. Multilineage cells from human adipose tissue: implications for cell-based therapies. *Tissue Eng.* 2001;7(2):211–228.

[48] Rangappa S, Fen C, Lee EH, Bongso A, Wei ESK. Transformation of adult mesenchymal stem cells isolated from the fatty tissue into cardiomyocytes. *Ann Thor Surg.* 2003;75:775–779.

[49] Tholpady SS, Katz AJ, Ogle RC. Mesenchymal stem cells from rat visceral fat exhibit multipotential differentiation *in vitro. Anat Rec.* 2003;272a:398–402.

[50] Gronthos S, Franklin DM, Leddy HA, Robey PG, Storms RW, Gimble JM. Surface protein characterization of human adipose tissue-derived stromal cells. *J Cell Physiol.* 2001;189(1):54–63.

[51] Zuk PA, Zhu M, Ashjian P, et al. Human adipose tissue is a source of multipotent stem cells. *Mol Biol Cell.* 2002;13(12):4279–4295.

[52] Noth U, Osyczka AM, Tuli R, Hickok NJ, Danielson KG, Tuan RS. Multilineage mesenchymal differentiation potential of human trabecular bone-derived cells. *J Orthop Res.* 2002;20(5):1060–1069.

[53] Zohar R, Sodek J, McCulloch CAG. Characterization of stromal progenitor cells enriched by flow cytometry. *Blood.* 1997;90(9):3471–3481.

[54] Grigoriadis AE, Heersche JN, Aubin JE. Differentiation of muscle, fat, cartilage, and bone from pro-genitor cells present in a bone-derived clonal cell population: effect of dexamethasone. *J Cell Biol.* 1988;106:2139–2151.

[55] Gronthos S, Zannettino AC, Graves SE, Ohta S, Hay SJ, Simmons PJ. Differential cell surface expression of the STRO-1 and alkaline phosphatase antigens on discrete developmental stages in primary cultures of human bone cells. *J Bone Miner Res.* 1999;14:47–56.

[56] Williams JT, Southerland SS, Souza J, Calcutt AF, Cartledge RG. Cells isolated from adult human skele-tal muscle capable of differentiating into multiple mesodermal phenotypes. *Am Surg.* 1999;65(1):22–26.

[57] Beauchamp JR, Morgan JE, Pagel CN, Partridge TA. Dynamics of myoblast transplantation reveal a discrete minority of precursors with stem cell-like properties as the myogenic source. *J Cell Biol.* 1999;144:1113–1122.

[58] Lipton BH, Schultz E. Developmental fate of skeletal muscle satellite cells. *Science.* 1979;205:1292–1294.

[59] Warejcka DJ, Harvey R, Taylor BJ, Young HE, Lucas PA. A population of cells isolated from rat heart capable of differentiating into several mesodermal phenotypes. *J Surg Res.* 1996;62:233–242.

[60] Tallheden T, Dennis JE, Lennon DP, Sjögren-Jansson, Caplan AI, Lindahl A. Phenotypic plasticity of human articular chondrocytes. *J Bone Joint Surg.* 2003;85A, S2, 93–100.

[61] Bernard-Beaubois K, Hecquet C, Houcine O, Hayem G, Adolphe M. Culture and characterization of juvenile rabbit tenocytes. *Cell Biol Toxicol.* 1997;13(2):103–113.

[62] Charbord P, Oostendorp R, Pang W, et al. Compartative study of stromal cell lines derived from embryonic, fetal, and postnatal mouse blood-forming tissues. *Exp Hematol.* 2002;30:1202–1210.

[63] Galmiche MC, Koteliansky VE, Briere J, Herve P, Charbord P. Stromal cells from human long-term marrow cultures are mesenchymal cells that differentiate following a vascular smooth muscle differentiation pathway. *Blood.* 1993;82:66–76.

[64] Gronthos S, Mankani M, Brahim J, Gehron Robey P, Shi S. Postnatal human dental pulp stem cells (DPSCs) *in vitro* and *in vivo. Proc Natl Acad Sci USA.* 2000;97(25):13625–13630.

[65] Almeida-Porada G, El Shabrawy D, Porada C, Zanjani ED. Differentiative potential of human metanephric mesenchymal cells. *Exp Hematol.* 2002;30:1454–1462.

[66] Owen M, Friedenstein AJ. Stromal stem cells: marrow derived osteogenic precursors. *Ciba Foundat Symp.* 1988;136:42–60.

[67] Friedenstein AJ, Chailakhyan RK, Gerasimov UV. Bone marrow osteogenic stem cells: *in vitro* cultivation and transplantation in diffusion chambers. *Cell Tissue Kinet.* 1987;20(3):263–272.

[68] Castro-Malaspina H, Gay RE, Resnick G, et al. Characterization of human bone marrow fibroblast colony-forming cells (CFU-F) and their progeny. *Blood.* 1980;56(2):289–301.

[69] Friedenstein AJ, Deriglasova UF, Kulagina NN, et al. Precursors for fibroblasts in different populations of hematopoietic cells as detected by the *in vitro* colony assay method. *Exp Hematol.* 1974;2(2):83–92.

[70] Bruder SP, Jaiswal N, Haynesworth SE. Growth kinetics, self-renewal, and the osteogenic potential of purified human mesenchymal stem cells during extensive subcultivation and following cryopreservation. *J Cell Biochem.* 1997;64:278–294.

[71] Phinney DG, Kopen G, Righter W, Webster S, Tremain N, Prockop DJ. Donor variation in the growth properties and osteogenic potential of human marrow stromal cells. *J Cell Biochem.* 1999;75:424–436.

[72] DiGirolamo C, Stokes D, Colter D, Phinney DG, Class R, Prockop DJ. Propagation and senescence of human marrow stromal cells in culture: a simple colony-forming assay identifies samples with the greatest potential to propagate and differentiate. *Br J Haemat.* 1999;107:275–281.

[73] Koç ON, Peters C, Aubourg P, et al. Bone marrow-derived mesenchymal stem cells remain host-derived despite successful hematopoietic engraftment after allogeneic transplantation in patients with lysosomal and peroxismal storage diseases. *Exp Hematol.* 1999;27:1675–1681.

[74] Murphy JM, Dixon K, Beck S, Fabian D, Feldman A, Barry F. Reduced chondrogenic and adipogenic activity of mesenchymal stem cells from patients with advanced osteoarthritis. *Arthr Rheum.* 2002;46(3):704–713.

[75] Galotto M, Berisso G, Delfino L, et al. Stromal damage as consequence of high-dose chemo/radiotherapy in bone marrow transplant recipients. *Exp Hematol.* 1999;27:1460–1466.

[76] Satomura K, Derubeis AR, Fedarko NS, et al. Receptor tyrosine kinase expression in human bone marrow stromal cells. *J Cell Physiol.* 1998;177(3):426–438.

[77] Gronthos S, Simmons PJ. The Growth factor requirements of STRO-1-positive human bone marrow stromal precursors under serum-deprived conditions *in vitro. Blood.* 1995;85(4):929–940.

[78] Lennon DP, Haynesworth SE, Young RG, Dennis JE, Caplan AI. A chemically defined medium supports *in vitro* proliferation and maintains the osteochondral potential of rat marrow-derived mesenchymal stem cells. *Exp Cell Res.* 1995;219:211–222.

[79] Dartsch PC, Weiss HD, Betz E. Human vascular smooth muscle cells in culture: growth characteristics and protein pattern by use of serum-free media supplements. *Eur J Cell Biol.* 1990;51(2):285–294.

[80] Sekiya I, Larson BL, Smith JR, Pochampally R, Cui JG, Prockop DJ. Expansion of human adult stem cells from bone marrow stroma: conditions that maximize the yields of early progenitors and evaluate their quality. *Stem Cells.* 2002;6:530–541.

[81] Gronthos S, Graves SE, Simmons PJ. Isolation, purification and *in vitro* manipulation of human bone marrow stromal precursor cells. In: Beresford JN and Owen ME, eds. *Marrow Stromal Cell Culture.* United Kingdom: Cambridge University Press; 1998:chap 3.

[82] Pereira RF, Halford KW, O'Hara MD, et al. Cultured adherent cells from marrow serve as long-lasting precursor cells for bone, cartilage, and lung in irradiated mice. *Proc Natl Acad Sci USA.* 1995;92:4857–4861.

[83] Horwitz EM, Gordon PL, Koo WKK, et al. Isolated allogeneic bone marrow-derived mesenchymal cells engraft and stimulate growth in children with osteogenesis imperfecta: implications for cell therapy of bone. *Proc Natl Acad Sci USA.* 2002;99(13):8932–8937.

[84] Majumdar MK, Banks V, Peluso DP, Morris EA. Isolation, characterization, and chondrogenic potential of human bone marrow-derived multipotential stromal cells. *J Cell Physiol.* 2000;185:98–106.

[85] Young RG, Butler DL, Weber W, Caplan AI, Gordon SL, Fink DJ. Use of mesenchymal stem cells in a collagen matrix for Achilles tendon repair. *J Ortho Res.* 1998;16:406–413.

[86] Kopen GC, Prockop DJ, Phinney DG. Marrow stromal cells migrate throughout forebrain and cerebellum, and they differentiate into astrocytes after injection into neonatal mouse brains. *Proc Natl Acad Sci USA.* 1999;96:10711–10716.

[87] Azizi SA, Stokes D, Augelli BJ, DiGirolamo C, Prockop DJ. Engraftment and migration of human bone marrow stromal cells implanted in the brains of albino rats-similarities to astrocyte grafts. *Proc Natl Acad Sci USA.* 1998;95:3908–3913.

[88] Dezawa M, Takahashi I, Esaka M, Takano M, Sawada H. Sciatic nerve regeneration inh rats induced by transplantation of *in vitro* differentiated bone-marrow stromal cells. *Eur J Neurosci.* 2001;14:1771–1776.

[89] Woodbury D, Reynolds K, Black IB. Adult bone marrow stromal stem cells express germline, ectodermal, endodermal, and mesodermal genes prior to neurogenesis. *J Neurosci Res.* 2002;96:908–917.

[90] Ferrari G, Cusella-De Angelis G, Coletta M, et al. Muscle regeneration by bone marrow-derived myogenic progenitors. *Science.* 1998;279:1528–1530.

[91] Toma C, Pittenger MF, Cahill KS, Byrne BJ, Kessler PD. Human mesenchymal stem cells differentiate to a cardiomyocyte phenotype in the adult murine heart. *Circulation.* 2002;105:93–98.

[92] Shake JG, Gruber PJ, Baumgartner WA, et al. Mesenchymal stem cell implantation in a swine myocardial infarct model: engraftment and functional effects. *Ann Thorac Surg.* 2002;73:1919–1926.

[93] Hakuno D, Fukuda K, Makino S, et al. Bone marrow-derived regenerated cardiomyocytes (CMG cells) express functional adrenergic and muscarinic receptors. *Circulation.* 2002;105:380–386.

[94] Wang JS, Shum-Tim D, Galipeau J, Chedrawy E, Eliopoulos N, Chiu RC. Marrow stromal cells for cellular cardiomyoplasty: feasibility and potential clinical advantages. *J Thorac Cardiovasc Surg.* 2000;120(5):999–1005.

[95] Jaiswal N, Haynesworth SE, Caplan AI, Bruder SP. Osteogenic differentiation potential of purified, culture-expanded human mesenchymal stem cells *in vitro. J Cell Biochem.* 1997;64:295.

[96] Muraglia A, Cancedda R, Quarto R. Clonal mesenchymal progenitors from human bone marrow differentiate *in vitro* according to a hierarchical model. *J Cell Sci.* 2000;113:1161–1166.

[97] Gronthos S, Graves SE, Ohta S, Simmons PJ. The STRO-1+ fraction of adult human bone marrow contains the osteogenic precursors. *Blood.* 1994;84(12):4164–4173.

[98] Dennis JE, Carbillet JP, Caplan AI, Charbord P. The STRO-1+ marrow cell population is multipotential. *Cells Tissues Organs.* 2002;170:73–82.

[99] Simmons PJ, Torok-Storb B. Identification of stromal cell precursors in human bone marrow by a novel monoclonal antibody, STRO-1. *Blood.* 1991;78(1):55–62.

[100] Caplan AI. Mesenchymal stem cells. *J Orthop Res.* 1991;9(5):641–650.

[101] Johnstone B, Hering TM, Caplan AI, Goldberg VM, Yoo JU. *In vitro* chondrogenesis of bone marrow derived mesenchymal progenitor cells. *Exp Cell Res.* 1998;238:265–272.

[102] Petersen BE, Bowen WC, Patrene KD, et al. Bone marrow as a potential source of hepatic oval cells. *Science.* 1999;284:1168–1170.

[103] Theise ND, Badve S, Saxena R, et al. Derivation of hepatocytes from bone marrow cells in mice after radiation-induced myeloablation. *Hepatology.* 2000;31:235–240.

[104] Theise ND, Nimmakayalu M, Gardner R, et al. Liver from bone marrow in humans. *Hepatology.* 2000;32:11–16.

[105] Brazelton TR, Rossi FMV, Keshet GI, Blau HM. From marrow to brain: expression of neuronal phenotypes in adult mice. *Science.* 2000;290:1775–1779.

[106] Mezey É, Chandross KJ, Harta G, Maki RA, McKercher SR. Turning blood into brain: cells bearing neuronal antigens generated *in vivo* from bone marrow. *Science.* 2000;290:1779–1782.

[107] Bjornson CRR, Rietze RL, Reynolds BA, Magli MC, Vescovi AL. Turning brain into blood: a hematopoietic fate adopted by adult neural stem cells *in vivo. Science.* 1999;283:534–537.

[108] Dennis JE, Charbord P. Origin and differentiation of human and murine stroma. *Stem Cells.* 2002;20:205–214.

[109] Woodbury D, Schwarz EJ, Prockop DJ, Black IB. Adult rat and human bone marrow stromal cells differentiate into neurons. *J Neurosci Res.* 2000;61:364–370.

[110] Hu M, Krause D, Greaves M, et al. Multilineage gene expression precedes commitment in the hemopoietic system. *Genes Dev.* 1997;774–785.

[111] Cross MA, Enver T. The lineage commitment of haemapoietic progenitor cells. *Curr Opin Genet Dev.* 1997;7:609–613.

[112] Seshi B, Kumar S, Sellers D. Human bone marrow stromal cell, coexpression of markers specific for multiple mesenchymal cell lineages. *Blood Cell Mol Dis.* 2000;26:234–246.

[113] Tagami M, Ichinose S, Yamagata K, et al. Genetic and ultrastructural demonstration of strong reversibility in human mesenchymal stem cell. *Cell Tissue Res.* 2003;312:31–40.

[114] Sanchez-Ramos J, Song S, Cardozo-Pelaez F, et al. Adult bone marrow stromal cells differentiate into neural cells *in vitro. Exp Neurol.* 2000;164(2):247–56.

[115] Zhao LR, Duan WM, Reyes M, Keene CD, Verfaillie CM, Low WC. Human bone marrow stem cells exhibit neural phenotypes and ameliorate neurological deficits after grafting into the ischemic brain of rats. *Exp Neurol.* 2002;174(1):11–20.

[116] Wakitani S, Saito T, Caplan AI. Myogenic cells derived from rat bone marrow mesenchymal stem cells exposed to 5-azacytidine. *Muscle Nerve.* 1995;18:1417–1426.

[117] Makino S, Fukuda K, Miyoshi S, et al. Cardiomyocytes can be generated from marrow stromal cells *in vitro. J Clin Invest.* 1999;103(5):697–705.

[118] Tomita S, Li R, Weisel R, et al. Autologous transplantation of bone marrow cells improves damaged heart function. *Circulation.* 1999;100:II247–II256.

[119] Fukuda K. Reprogramming of bone marrow mesenchymal stem cells into cardiomyocytes. *Crit Rev Biol.* 2002;325(10):1027–1038.

[120] Reyes M, Lund T, Lenvik T, Aguiar D, Koodie L, Verfaillie CM. Purification and *ex vivo* expansion of postnatal human marrow mesodermal progenitor cells. *Blood*. 2001;98:2615–2625.

[121] Jiang Y, Jahagirdar BN, Reinhardt RL, et al. Pluripotency of mesenchymal stem cells derived from adult marrow. *Nature*. 2002, advance online publication.

[122] Colter DC, Sekiya I, Prockop DJ. Identification of a subpopulation of rapidly self-renewing and multipotential adult stem cells in colonies of human marrow stromal cells. *Proc Natl Acad Sci USA*. 2001;98(4):7841–7845.

[123] Colter DC, Class R, DiGirolamo CM, Prockop DJ. Rapid expansion of recycling stem cells in cultures of plastic-adherent cells from human bone marrow. *Proc Natl Acad Sci*. 2000;97(7):3213–3218.

[124] Simmons PJ, Torok-Storb B. CD34 expression by stromal precursors in normal human adult bone marrow. *Blood*. 1991;78:2848.

[125] Haynesworth SE, Baber MA, Caplan AI. Cell surface antigens on human marrow-derived mesenchymal cells are detected by monoclonal antibodies. *Bone*. 1992;13(1):69–80.

[126] Barry FP, Boynton RE, Haynesworth S, Murphy JM, Zaia J. The monoclonal antibody SH-2, raised against human mesenchymal stem cells, recognizes an epitope on endoglin (CD105). *Biochem Biophys Res Commun*. 1999;265:134–139.

[127] Barry F, Boynton RE, Murphy M, Zaia J. The SH-3 and SH-4 antibodies recognize distinct epitopes on CD73 from human mesenchymal stem cells. *Biochem Biophys Res Commun*. 2001;289:519–524.

[128] *Seventh International Workshop on Human Leukocyte Differentiation Antigens*. 2003.

[129] Bruder SP, Ricalton NS, Boynton RE, et al. Mesenchymal stem cell surface antigen SB-10 corresponds to activated leukocyte-cell adhesion molecule and is involved in osteogenic differentiation. *J Bone Miner Res*. 1998;13(4):655–663.

[130] Chichester CO, Fernandez M, Minguell JJ. Extracellular matrix gene expression by human bone marrow stroma and by marrow fibroblasts. *Cell Adhes Commun*. 1993;1(2):93–99.

[131] Andrades JA, Nimni ME, Han B, Ertl DC, Hall FL, Becerra J. Type I collagen combined with a recombinant TGF-beta serves as a scaffold for mesenchymal stem cells. *Int J Dev Biol*. 1996;S1:1073.

[132] Robledo MM, Hidalgo A, Lastres P, et al. Characterization of TGF-beta 1-binding proteins in human bone marrow stromal cells. *Br J Haematol*. 1996;93(3):507–514.

[133] Gordon EM, Skotzko M, Kundu RK, et al. Capture and expansion of bone marrow-derived mesenchymal progenitor cells with a transforming growth factor-beta1-von Willebrand's factor fusion protein for retrovirus-mediated delivery of coagulation factor IX. *Hum Gene Ther*. 1997;8(11):1385–1394.

[134] Quirici N, Soligo D, Bossolasco P, Servida F, Lumini C, Lambertenghi Deliliers G. Isolation of bone marrow mesenchymal stem cells by anti-nerve growth factor receptor antibodies. *Exp Hematol*. 2002;30:783–791.

[135] Guerriero A, Worford L, Holland HK, Guo GR, Sheehan K, Waller EK. Thrombopoietin is synthesized by bone marrow stromal cells. *Blood*. 1997;90(9):3444–3455.

[136] Simmons PJ, Gronthos S, Zannettino ACW. The development of stromal cells. In: Son L, ed. *Hematopoiesis: A developmental approach*. New York: Oxford University Press; 2001:718–726.

[137] Deschaseaux F, Charbord P. Human marrow stromal precursors are alpha1 integrin subunit-positive.*J Cell Physiol*. 2000;184:319–325.

[138] Filshie RJA, Zannettino ACW, Makrynikola V. MUC18, a member of the immunoglobulin superfamily, is expressed on bone marrow fibroblasts and a subset of haematological malignancies. *Leukemia*. 1998;12:414–421.

[139] Simmons PJ, Gronthos S, Zannettino A, Ohta S, Graves S. Isolation, characterization and functional activity of human marrow stromal progenitors in hemopoiesis. *Prog Clin Biol Res*. 1994;389:271–280.

[140] Gronthos S, Franklin DM, Leddy HA, Gehron-Robey P, Storms RW, Gimble JM. Surface protein characterization of human adipose tissue-derived stromal cells. *J Cell Physiol*. 2001;189:54–63.

[141] Wineman J, Moore K, Lemischka I, Muller-Sieburg C. Functional heterogeneity of the hematopoietic microenvironment: rare stromal elements maintain long-term repopulating cells. *Blood*. 1996;87(10):4082–4090.

[142] Gualtieri RJ, Liang CM, Shadduck RK, Waheed A, Banks J. Identification of the hematopoietic growth factors elaborated by bone marrow stromal cells using antibody neutralization analysis. *Exp Hematol*. 1987;15:883–889.

[143] Majumdar MK, Thiede MA, Haynesworth SE, Bruder SP, Gerson SL. Human marrow-derived mesenchymal stem cells (MSCs) express hematopoietic cytokines and support long-term hematopoiesis when differentiated toward stromal and osteogenic lineages. *J Hematother Stem Cell Res*. 2000;9(6):841–848.

[144] Haynesworth SE, Baber MA, Caplan AI. Cytokine expression by human marrow-derived mesenchymal progenitor cells *in vitro*: effects of dexamethasone and IL-1 alpha. *J Cell Physiol*. 1996;166(3):585–592.

[145] Thalmeier K, Meissner P, Reisbach G, et al. Constitutive and modulated cytokine expression in two permanent human bone marrow stromal cell lines. *Exp Hematol*. 1996;24(1):1–10.

[146] Brouard N, Chapel A, Thierry D, Charbord P, Peault B. Transplantation of gene-modified human bone marrow stromal cells into mouse-human bone chimeras. *J Hematother Stem Cell Res*. 2000;9(2):175–181.

[147] Ding L, Lu S, Batchu R, III RS, Munshi N. Bone marrow stromal cells as a vehicle for gene transfer. *Gene Ther*. 1999;6(9):1611–1616.

[148] Caplan AI, Bruder SP. Mesenchymal stem cells: building blocks for molecular medicine in the 21st century. *Trends Mol Med*. 2001;7(6):259–264.

[149] Li KJ, Dilber MS, Abedi MR, et al. Retroviral-mediated gene transfer into human bone marrow stromal cells: studies of efficiency and *in vivo* survival in SCID mice. *Eur J Haematol*. 1995;55(5):302–306.

[150] Nolta JA, Hanley MB, Kohn DB. Sustained human hematopoiesis in immunodeficient mice by cotransplantation of marrow stroma expressing human interleukin-3: analysis of gene transduction of long-lived progenitors. *Blood*. 1994;83(10):3041–3051.

[151] Mosca JD, Hendricks JK, Buyaner D, et al. Mesenchymal stem cells as vehicles for gene delivery. *Clin Orthop Rel Res*. 2000;379S:S71–S90.

[152] Allay JA, Dennis JE, Haynesworth SE, et al. LacZ and IL-3 expression *in vivo* after retroviral transduction of marrow-derived human osteogenic mesenchymal progenitors. *Hum Gene Ther*. 1997;8(12):1417–1427.

[153] Lee K, Wang G, Buyaner D. Retroviral transduced human mesenchymal stem cells: maintenance of expression and efficacy during expansion and lineage differentiation. *Exp Hematol*. 1999;27:53.

[154] Marx JC, Allay JA, Persons DA, et al. High-efficiency transduction and long-term gene expression with a murine stem cell retroviral vector encoding the green fluorescent protein in human marrow stromal cells. *Hum Gene Ther*. 1999;10(7):1163–1173.

[155] Lee K, Majumdar MK, Buyaner D, Hendricks JK, Pittenger MF, Mosca JD. Human mesenchymal stem cells maintain transgene expression during expansion and differentiation. *Mol Ther*. 2001;3(6):857–866.

[156] Oyama M, Tatlock A, Fukuta S, et al. Retrovirally transduced bone marrow stromal cells isolated from a mouse model of human osteogenesis imperfecta (oim) persist in bone and retain the ability to form cartilage and bone after extended passaging. *Gene Ther*. 1999;6(3):321–329.

[157] Bulabois CE, Yerly-Motta V, Mortensen BT, et al. Retroviral-mediated marker gene transfer in hematopoiesis-supportive marrow stromal cells. *J Hematother*. 1998;7(3):225–239.

[158] Suzuki K, Oyama M, Faulcon L, Robbins PD, Niyibizi C. *In vivo* expression of human growth hormone by genetically modified murine bone marrow stromal cells and its effect on the cells *in vitro*. *Cell Transplant*. 2000;9(3):319–327.

[159] Bartholomew A, Patil S, Mackay A, et al. Baboon mesenchymal stem cells can be genetically modified to secrete human erythropoietin *in vivo*. *Hum Gene Ther*. 2001;12(12):1527–1541.

[160] Dao MA, Pepper KA, Nolta JA. Long-term cytokine production from engineered primary human stromal cells influences human hematopoiesis in an *in vivo* xenograft model. *Stem Cells*. 1997;15:443–454.

[161] Conget PA, Minguell JJ. Adenoviral-mediated gene transfer into *ex vivo* expanded human bone marrow mesenchymal progenitor cells. *Exp Hematol*. 2000;28(4):382–390.

[162] Gazit D, Turgeman G, Kelley P, et al. Engineered pluripotent mesenchymal cells integrate and differentiate in regenerating bone: a novel cell-mediated gene therapy. *J Gene Med*. 1999;1(2):121–133.

[163] Cancedda R, Dozin B, Giannoni P, Quarto R. Tissue engineering and cell therapy of cartilage and bone. *Matrix Biol*. 2003;22:81–91.

[164] Tuan RS, Boland G, Tuli R. Adult mesenchymal stem cells and cell-based tissue engineering. *Arthr Res Ther*. 2002;5(1):32–45.

[165] Pereira RF, O'Hara MD, Laptev AV, et al. Marrow stromal cells as a source of progenitor cells for nonhematopoietic tissues in transgenic mice with a phenotype of osteogenesis imperfecta. *Proc Natl Acad Sci USA*. 1998;95:1142–1147.

[166] Horwitz EM, Prockop DJ, Fitzpatrick LA, et al. Transplantability and therapeutic effects of bone marrow-derived mesenchymal cells in children with osteogenesis imperfecta. *Nat Med*. 1999;5(3):309–313.

[167] Saito T, Dennis JE, Lennon DP, Young RG, Caplan AI. Myogenic expression of mesenchymal stem cells within myotubes of mdx mice *in vitro* and *in vivo*. *Tissue Eng*. 1996;1:327–344.

[168] Keating A, Guinn B, Laraya P, Wang XH. Human marrow stromal cells electrotransfected with human factor IX (FIX) cDNA engraft in SCID mouse marrow and transcribe human FIX. *Exp Hematol.* 1996;24(9):S180.

[169] Hurwitz DR, Kirchgesser M, Merrill W, et al. Systemic delivery of human growth hormone or human factor IX in dogs by reintroduced genetically modified autologous bone marrow stromal cells. *Hum Gene Ther.* 1997;8(2):137–156.

[170] Chuah MK, Van Damme A, Zwinnen H, et al. Long-term persistence of human bone marrow stromal cells transduced with factor VIII-retroviral vectors and transient production of therapeutic levels of human factor VIII in nonmyeloablated immunodeficient mice. *Hum Gene Ther.* 2000;11(5):729–738.

[171] Turgeman G, Pittman DD, Muller R, et al. Engineered human mesenchymal stem cells: a novel platform for skeletal cell mediated gene therapy. *J Gene Med.* 2001;3(3):240–251.

[172] Campagnoli C, Bellantuono I, Kumar S, Fairbairn LJ, Roberts I, Fisk NM. High transduction efficiency of circulating first trimester fetal mesenchymal stem cells: potential targets for in utero *ex vivo* gene therapy. *BJOG.* 2002;109(8):952–954.

[173] Van Damme A, Vanden Driessche T, Collen D, Chuah MK. Bone marrow stromal cells as targets for gene therapy. *Curr Gene Ther.* 2002;2(2):195–209.

[174] Laver J, Jhanwar SC, O'Reilly RJ, Castro-Malaspina H. Host origin of the human hematopoietic microenvironment following allogeneic bone marrow transplantation. *Blood.* 1987;70(6):1966–1968.

[175] Simmons PJ, Przepiorka D, Donnall Thomas E, Torok-Storb B. Host origin of marrow stromal cells following allogeneic bone marrow transplantation. *Nature.* 1987;328:429–432.

[176] Agematsu K, Nakahori Y. Recipient origin of bone-marrow-derived fibroblastic stromal cells during all periods following bone marrow transplantation in humans. *Br J Haematol.* 1991;79:359–365.

[177] Santucci MA, Trabetti E, Martinelli G, et al. Host origin of bone marrow fibroblasts following allogeneic bone marrow transplantation for chronic myeloid leukemia. *Bone Marrow Transplant.* 1992;10:255–259.

[178] Awaya N, Rupert K, Bryant E, Torok-Storb B. Failure of adult marrow-derived stem cells to generate marrow stroma after successful hematopoietic stem cell transplantation. *Exp Hematol.* 2002;30:937–942.

[179] Galotto M, Berisso G, Delfino L, et al. Stromal damage as consequence of high-dose chemo/radiotherapy in bone marrow transplant recipients. *Exp Hematol.* 1999;27:1460–1466.

[180] Lazarus HM, Haynesworth SE, Gerson SL, Rosenthal NS, Caplan AI. Ex vivo expansion and subsequent infusion of human bone marrow-derived stromal progenitor cells (mesenchymal progenitor cells): implications for therapeutic use. *Bone Marrow Transplant.* 1995;16:557–564.

[181] Koç ON, Gerson SL, Cooper BW, et al. Rapid hematopoietic recovery after coinfusion of autologous-blood stem cells and culture-expanded marrow mesenchymal stem cells in advanced breast cancer patients receiving high-dose chemotherapy. *J Clin Oncol.* 2000;18:307–316.

[182] Gao J, Dennis JE, Muzic RF, Lundberg M, Caplan AI. The dynamic in vivo distribution of bone marrow-derived mesenchymal stem cells after infusion. *Cells Tissues Organs.* 2001;169(1):12–20.

[183] Devine S, Bartholomew AM, Mahmud N, et al. Mesenchymal stem cells are capable of homing to the bone marrow of non-human primates following systemic infusion. *Exp Hematol.* 2001;29:244–255.

[184] Bartholomew A, Sturgeon C, Siatskas M, et al. Mesenchymal stem cells suppress lymphocyte proliferation *in vitro* and prolong skin graft survival *in vivo*. *Exp Hematol.* 2002;30:42–48.

[185] Di Nicola M, Carlo-Stella C, Magni M, et al. Human bone marrow stromal cells suppress T-lymphocyte proliferation induced by cellular on nonspecific mitogenic stimuli. *Blood.* 2002;99:3838–3843.

[186] Le Blanc K, Tammik L, Sundberg B, Haynesworth SE, Ringdén O. Mesenchymal stem cells inhibit and stimulate mixed lymphocyte cultures and mitogenic responses independently of the major histocompatibility complex. *Scand J Immunol.* 2003;57(1):11–20.

[187] Jorgensen C, Djouad F, Apparailly F, Noël D. Engineering mesenchymal stem cells for immunotherapy. *Gene therapy.* 2003;10:928–931.

[188] Koç ON, Day J, Nieder M, Gerson SL, Lazarus HM, Krivit W. Allogeneic mesenchymal stem cell infusion for treatment of metachromatic leukodystrophy (MLD) and Hurler syndrome (MPS-IH). *Bone Marrow Transplant.* 2002;30:215–222.

[189] Lee ST, Jang JH, Cheong JW, et al. Treatment of high risk acute myelogenous leukaemia by myeloablative chemoradiotherapy followed by co-infusion of T-cell depleted haematopoietic stem cells and culture-expanded marrow mesenchymal stem cells from a related donor with one fully mismatched human leukocyte antigen haplotype. *Br J Haematol.* 2002;118:1128–1131.

[190] Fu SC, Wong YP, Chan BP, et al. The roles of bone morphogenetic protein (BMP) 12 in stimulating the proliferation and matrix production of human patellar tendon fibroblasts. *Life Sci.* 2003;72(26):2965–2974.

[191] Phinney DG, Kopen G, Isaacson RI, Prockop DJ. Plastic adherent stromal cells from the bone marrow of commonly used strains of inbred mice: variations in yield, growth, and differentiation. *J Cell Biochem.* 1999;72:570–585.

[192] Bianchi G, Muraglia A, Daga A, Corte G, Cancedda R, Quarto R. Microenvironment and stem properties of bone marrow-derived mesenchymal cells. *Wound Rep Reg.* 2001;9:460–466.

CHAPTER 2

ESTABLISHMENT AND TRANSDUCTION OF PRIMARY HUMAN STROMAL/MESENCHYMAL STEM CELL MONOLAYERS

T. MEYERROSE, I. ROSOVA, M. DAO, P. HERRBRICH, G. BAUER AND J. NOLTA

Washington University School of Medicine
Department of Internal Medicine, St Louis, MO

1. OVERVIEW

1.1. Introduction to mesenchymal stem cells for genetic engineering and cellular therapy

The rapidly dividing adherent myofibroblastic cells from human bone marrow are easily transduced using retroviral vectors. These cells were previously referred to as "stroma," or "marrow stromal cells (MSC)" but the more accurate term "Mesenchymal Stem Cells (MSC)" reflects the capacity of at least a subset of the population to differentiate into multiple tissues [1]. Our group and our collaborators have shown that human adipose and bone marrow-derived MSC are excellent vehicles from which to secrete proteins encoded by introduced transgenes *in vivo*, and will retain the capacity to differentiate into muscle, fat, fibroblast, cartilage, and to generate bone after transduction by viral vectors [2–4]. Stroma is initially a highly heterogenous cell mixture, consisting of fibroblasts, endothelial cells, adipocytes and macrophages. The current protocol, adapted from that which we originally developed in 1988, is optimized to rapidly and easily expand the multipotent, myofibroblastic component which is the cell of interest, commonly referred to as an "MSC" culture. In lieu of a facile sorting strategy, we and others have grown the cells out of marrow samples based on their ability to adhere to plastic and to rapidly expand in minimal medium. This chapter provides the instructions for expanding and transducing marrow stromal cell/mesenchymal stem cell monolayers.

Bone marrow stromal cells/mesenchymal stem cells are used for several purposes: (1) to engineer for sustained *in vivo* protein secretion, (2) to enhance retroviral-mediated gene transduction into hematopoietic stem cells, (3) as a feeder layer to maintain

J.A. Nolta (ed.), Genetic Engineering of Mesenchymal Stem Cells, 45–58.

hematopoietic cells in long-term bone marrow culture, (4) to study mesenchymal stem cell biology, and finally, (5) for tissue repair and various cellular therapies. MSC are being tested in early clinical trials to examine their potential to promote the repair of not only bone and cartilage, but also skeletal and cardiac muscle defects, liver, pancreas, and brain injury, and to enhance revascularization in multiple tissues (reviewed in [5–9] and discussed elsewhere in this volume). In addition, bone marrow-derived MSC have been reported to facilitate engraftment [10–14], a phenomenon which we have studied in immune deficient mice over the past decade [15–18].They have also been reported to reduce the incidence of graft vs. host disease, through their immune-suppressive qualities, potentially by the release of hepatocyte growth factor (HGF) as well as TGFβ, prostaglandins, and immunosuppressive levels of the enzyme in-doleamine 2,3-dioxygenase (IDO) [19–24]. Umbilical cord blood and adipose-derived MSC may have similar functions. Finally, bone marrow-derived MSC have been re-ported to home to areas of solid tumor revascularization [25, 26], and thus may be used as delivery vehicles to target ablative agents, such as herpes simplex virus thymidine kinase (HSV-TK), into dividing tumor cells. Ongoing research is avidly examining these potential uses of MSC, and this book was generated to provide methods and ideas to foster this research.

1.2. Mesenchymal stem cell isolation and transduction

True mesenchymal stem cells are likely a subset of the rapidly growing marrow stromal cell monolayer. However, the phenotype of the most primitive MSC compartment is not easily identified. Description of some of the markers that are found on MSC has been done. However, these markers may or may not be on overlapping subsets, and there has been no systematic analysis of the potentiality of differentially sorted populations, as has been done with human hematopoietic stem cells, using clonal analysis techniques [27–29]. As discussed elsewhere in this volume, there is no single marker to identify MSC [30]. Most commonly, a lack of CD45 expression in combination with markers such as CD105 and CD73 is used, in an attempt to purify MSC away from macrophages and endothelial progenitors in plastic-adherent monolayers [31–33]. This simple sort-ing strategy can be problematic, however. Human MSC are, indeed, negative for CD45; No MSC activity is found in the CD45+ population. This seems to be the only highly reproducible phenotypic characteristic so far, with these changing cells, which readily respond to microenvironmental cues, and share internal and membrane proteins with neighboring cells, as well as assimilating many proteins from surrounding cells. MSC are present in the adherent fraction after brief culture, but round up and enter the non-adherent fraction while dividing, so by starting the isolation with only the adherent layer, some expanding cells will be lost. CD73 is not MSC-specific, and the author's laboratory has discontinued the use of CD105 (endoglin) for MSC isolation, since it is also expressed on macrophage and endothelial cells [34–36], the primary contami-nants of adherent bone marrow and adipose-derived MSC monolayers. Endoglin is also present on murine hematopoietic stem cells [37, 38]. Anti-CD105/endoglin antibodies did not prove useful in subfractionating either human or murine MSC populations, in our hands. In our laboratory, the best strategy to date is to deplete BM samples for

CD45+ (hematopoietic), CD31+ (endothelial), and Glyco A+ (human erythroid pro-genitors) or ter119+ (murine erythroid progenitors) cells, to greatly enrich the MSC population in a fresh sample. This can be done using lineage depletion kits or flow-based cell sorting.

As well as a lack of a simple strategy using one MSC-specific marker to purify MSC away from hematopoietic and endothelial cells in a marrow or adipose sample, to date, a phenotypic hierarchy has not been established, so it is not known how to dissect the most primitive and pluripotent cells from the bulk population. Active research is ongoing in this area, and methods to allow the prospective isolation of primitive MSC, without the need for culture, will be important for future cellular therapy applications. We are currently using multiparameter sorting strategies in conjunction with viral marking to generate a clonotypic integration site, coupled with inverse PCR to identify primitive MSC capable of generating different lineages. This will allow prospective assessment of primitive, multipotential MSC phenotypes. We have described these techniques for hematopoietic stem cells [17, 27–29, 39]. Until the phenotype of the most primitive cells has been clearly defined, to allow isolation of undifferentiated cells directly from human bone marrow or adipose tissue, the field relies upon expanding cells in serum-containing medium, and then checking them for retention of the capacity to form bone, fat, cartilage, and muscle *in vitro*. More clinically relevant strategies are under development and are discussed in this volume.

2. OBTAINING AND PLATING HUMAN BONE MARROW-DERIVED MSC

2.1. Marrow filtration screens

The screens used to filter marrow during harvest are the richest source of mesenchymal stem cells, as we have described [3, 15]. Many small bony spicules packed with stroma (as well as hematopoietic stem and progenitor cells) will get lodged in the screen, and can be easily removed by flushing. The cells from one harvest screen, from a normal donor, should be split between 4 × T-75 vent-cap flasks in 15 mls of stromal medium (section 3) per flask. The cells will then be expanded, as described below (section 3). Filter vent-cap flasks are used for long-term culture, despite their greater cost as compared to standard screw-caps, because the risks for air-borne fungal spore contamination are high for cultures, which can be grown for 1–2 months. The tightly closed, gas permeable filter vent caps reduce the risk of cross-contamination between flasks.

Many transplant programs are now using G-CSF mobilized peripheral blood as a stem cell source in lieu of bone marrow. Unfortunately, MSC are not found in appreciable levels in G-CSF mobilized blood. The use of newer mobilization agents, such as the CXCR4 antagonist AMD3100 [40], might provide better MSC mobilization than the standard regimens. Until then, researchers at those institutions can purchase whole marrow, or even purified and cryopreserved MSC, from commercial sources, or might consider beginning a normal donor program for marrow donation to be used for research. In lieu of flushing harvest screens, when whole marrow is available from these sources, the investigator should proceed as directed below.

2.2. Bone marrow aspirates

If the richest source of MSC- bone marrow harvest screens- are not available, whole aspirated bone marrow can be used as a source of mesenchymal stem cells. Spicules from unseparated BM will be present in the aspirate, and can be collected by gravity sedimentation. The liquid marrow is then removed to another tube for additional processing. It is advisable to perform at least one red cell lysis and wash before plating, if using this method. The washing technique is described in section 2.3. The spicules from a 10 ml aspirate should then be plated in T-75 vent-cap flasks in 15 mls of stromal medium (Dexter's original medium = DOM, section 3), which is the richest medium and rapidly forces contaminating hematopoietic cells into erythroid and monocytic differentiation. A simpler medium can also be used, as described in section 3. If the aspirate providing the spicules is larger, the number of flasks should be scaled up accordingly. If the BM sample must be ficolled for other studies, and the MSC investigator is salvaging spicules, use the techniques described in the section below.

2.3. Spicules from RBC pellet of ficoll layer in marrow aspirate processing

The ficolled "buffy coat" is not a rich source of MSC, but is what many investigators have to work with. The MSC are far more rare in the aspirated marrow fraction than in spicules from the harvest screens. They are even rarer if the sample is first ficolled (approximately 1×10^6/ml in the "buffy coat" or mononuclear fraction. If whole marrow aspirates are to be used, an optimal strategy is to use the mononuclear fraction, and also to recover the spicules from the bottom of the 50 ml ficoll tubes, since the small pieces of bone will fall through the density gradient.

For ficolling, first mix an equal volume of whole marrow and 1x phosphate buffered saline and then gentl layer 25 mls over an equal volume of ficoll-paque in a 50 ml conical tube. Centrifuge the cells at 2000 rpm for 15 minutes. Once approximately 15 mls of the serum layer has been removed and discarded, buffy coat cells can be collected in another 10–15 mls, washed, and plated as described below. Then there are 5–10 mls of packed red blood cells and bony spicules left in the bottom of the tube. PBS should then be added up to a volume of 50 mls. Allow the tubes to settle upright for 3 minutes, without centrifugation. Remove 40 mls PBS and RBC, then repeat the washing step: add PBS, let the spicules settle out, and remove RBC/PBS down to the final 10 mls. At this point, the red blood cells are sufficiently diluted out to allow plating of the spicules. Add another 40 mls PBS, centrifuge, remove fluid down to the last ml, and plate as described below.

3. EXPANDING HUMAN MSC

3.1. Plating MSC (initial seeding)

If total (RBC lysed) or ficolled marrow is used, to expand mesenchymal stem cells, cells are plated at a concentration of 5×10^6 mononuclear cells per ml in 75 cm² flasks, in 15–20 mls total volume. Optimally, plate the cells in Dexter's original medium (DOM), which is prepared as shown in Table 2.1.A.

Table 2.1.A. Dexter's Original Medium for Stromal cells/MSC (DOM)

---

350 ml Iscove's Modified Dulbecco's Medium (IMDM)
75 ml heat- inactivated (HI) horse serum*
75 ml HI Fetal calf serum*
5 ml L-glutamine (200 mM stock)
2.5 ml Pen/Strep (stock = 10,000 U/ml penicillin and 10,000 ug/ml streptomycin)
500 ul 2-ME (10^{-1} M stock)
500 ul hydrocortisone (10^{-3} M stock)

Preparation of spicules for plating is described above, in sections 2.1 and 2.2. Spicules obtained from one harvest screen (section 2.1) should be divided between four T-75 flasks containing 15 mls of DOM or D10HG each. Spicules obtained from the RBC pellet resulting from 10–15 mls of ficolled marrow (section 2.3) can be plated in one T-75 flask, in 15 mls of medium. Cells are then expanded and transduced as described below.

3.2. MSC expansion

Following the initial seeding, described in section 3, the MSC are allowed to adhere to the flasks overnight. The next morning, non-adherent cells can be gently flushed from the flasks and replated in a second flask, in the same medium. The initial flask is refed fresh medium. DOM is the richest medium for MSC expansion without differentiation, and the horse serum rapidly forces contaminating hematopoietic cells into erythroid and monocytic differentiation, so that hematopoietic stem cells will not contaminate the stromal layer after three passages. A minimal medium, D10HG, which contains only fetal calf serum (Table 2.1.B), can also be used, but hematopoietic stem cells will survive happily on the stromal layer in this medium. MSC have not yet been expanded efficiently without the use of fetal calf serum, and it is imperative to screen the serum for optimal MSC growth without differentiation, when using either medium.

MSC colonies begin to develop as the cells expand out of the marrow spicules (Figures 2.1.A and 2.1.B, next pages). There are many other cells in the culture at this point. However, as the MSC grow and expand, the other cells differentiate out and/or can be removed. In the fetal calf/horse serum mixture (DOM-Table 2.1.A), the developing erythroid cells become non-adherent and are easily flushed away as the MSC layer develops and is expanded. Alternately, a depletion step can be done at passage 2–3, to

Table 2.1.B. Dulbecco's Modified Eagles medium with 10% fetal calf serum and high glucose (D10HG)

---

450 ml Dulbecco's Modified Eagles medium with high glucose
50 ml heat- inactivated (HI) Fetal calf serum*
5 ml L-glutamine (200 mM stock)
2.5 ml Pen/Strep (stock = 10,000 U/ml penicillin and 10,000 ug/ml streptomycin)

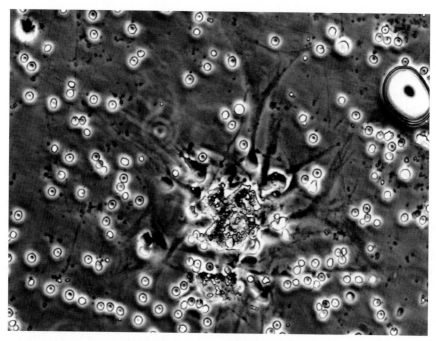

Figure 2.1.A. Human MSC beginning to grow out of a bony spicule obtained by gravity sedimentation from a normal donor bone marrow aspirate.

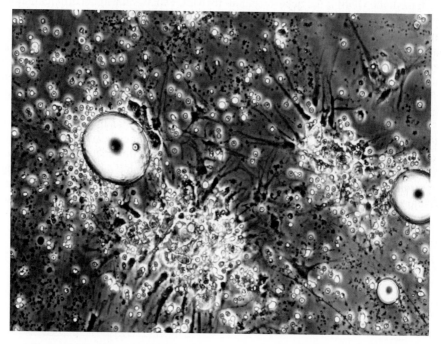

Figure 2.1.B. Human MSC beginning to grow out of spicules day 3 after plating.

remove Glycophorin A+ cells. Early monocytes can also be removed by flushing, but mature macrophages are tightly adherent to the tissue culture flask and cannot be removed, even with trypsin. Therefore, the MSC can be taken to a new flask, while leaving the macrophages behind to be discarded.

Alternately, the cells can be collected using an EDTA-based cell issociation buffer (rather than trypsin, which cleaves away many cell surface proteins), and then a FACS-based depletion can be done to remove CD45+ cells, including CD14+ mono-cyte/macrophages, from the developing MSC monolayer.

If not using FACS to fractionate MSC subpopulations based on cell surface proteins, it is best to use trypsinization (trypsin-EDTA), to dissociate sub-confluent monolayers of primary mesenchymal stem cells from the flask. For general maintenance and expansion, the cells should be "split" no more than 1:10 when they reach 70–80% confluency (Figure 2.2.A, following page).

Stroma is not usually transduced or used for other studies until passage #3 or 4. At this point (Figure 2.2.A), most hematopoietic cells will have been eliminated, except for mature macrophages, which typically will comprise less than 1% of the culture.

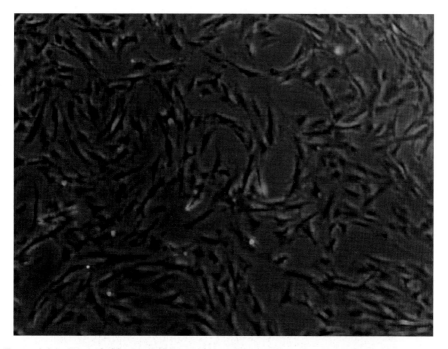

Figure 2.2A. Expanded human MSC at 70% confluency. This passage 3 culture was grown out of bone marrow spicules in Dexter's original medium (DOM). The cells have a fairy uniform myofibroblastic appearance.

The cells should be used for transduction, experiments, or transplantation between passage 3–6 for optimal results. By passage ten, they can begin to differentiate and become senescent. Since the primary MSC cultures are not immortalized, they do have a finite lifespan, and by later passage, they begin to slow down in growth and to become larger and more differentiated. At this point (passage 10) the cells will take on the appearance seen in Figure 2.2.B.

4. DETAILED METHODS

4.1. MSC isolation and expansion

1. Plate ficolled (buffy coat) marrow in D10HG (Table 2.1.B) at a concentration of 5×10^6 cells per ml, in 75 cm^2 flasks. Put the flasks into the incubator, at 37°C, with 5%

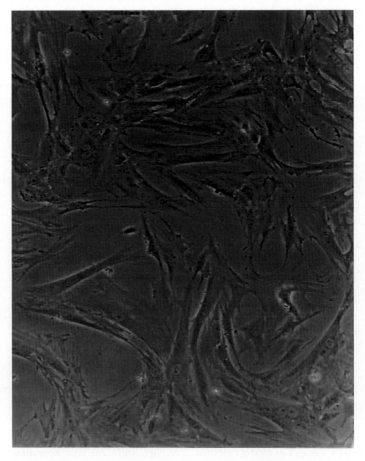

Figure 2.2B. Passage 10 human MSC that are senescent and differentiating. Using cells from this late passage would compromise experimental results.

CO_2. Alternately, plate spicules from harvest screens or from the RBC pellet of the mononuclear fraction, as described in sections 2.1 and 2.3, respectively.

2. 12–24 hours after plating, remove the nonadherent fraction, which contains primarily hematopoietic cells (can be used for HSC studies). Flush the adherent layer with PBS to remove as many hematopoietic cells as possible, and add the flushings to the collected nonadherent cell fraction. The nonadherent hematopoietic cells can be cryopreserved for later use if desired: the spicules are a rich source of hematopoietic, as well as mesenchymal stem cells. Refeed each adherent layer 15 mls of DOM (reagents section), for expansion of mesenchymal stem cells.

3. When the mesenchymal stem cells reach 70–80% confluency (70–80% of the plastic flask surface covered, Figure 2.2), split them by trypsinization. Remove the medium from the flask, and discard. Rinse the flask with 15 mls 1xPBS and discard. Add 2 mls trypsin/EDTA solution, and tip the flask back and forth gently, to completely coat the adherent layer. Remove excess trypsin, leaving approximately 500 uls in the flask.

4. Incubate for 10–15 minutes at 37°C. Pick up the flask and turn it to coat all surfaces every 3–4 minutes during the trypsinization process. Mesenchymal stem cells will be readily trypsinized if they are in subconfluent monolayers. If they are allowed to become confluent, they form a 3-dimensional tissue with excessive buildup of collagen and other extracellular matrix molecules between the layers of cells. The collagen layers are harder for the trypsin to digest than the adhesion foci with which the cells adhere to the plastic flask. The result is a useless sheet, or large chunks, of cells, which will quickly deplete nutrients from the medium, and will necrose in the center. If the mesenchymal stem cells were healthy and subconfluent prior to trypsinization, a single cell suspension will result.

5. To neutralize the trypsin, resuspend the cells from each flask in 45 mls DOM or other serum-containing medium. Transfer 15 mls each to three new flasks. Discard the original flask, which will contain firmly adherent macrophages.

6. Grow the mesenchymal stem cells until they reach 80% confluency, once again. Repeat steps 4–6. Grow up and repeat, to generate a "passage 3" layer. The monolayer should be a smooth, homogeneous mesenchymal stem cell population. The cells will be rather "chunky," not spindle shaped as will happen in straight FCS without the addition of horse serum. If the monolayer has numerous phase-bright macrophage contaminants, perform a CD45+ cell depletion using magnetic beads or FACS, or repeat steps 4–6. The resultant monolayer will be completely CD45-negative, due to the loss of hematopoietic cells. No phase-bright cells will be seen adhering to the MSC monolayer (Figure 2.2.A).

7. When the mesenchymal stem monolayer has the correct appearance (Figure 2.2.A), collect the cells from one 80% confluent flask containing passage 3–6 mesenchymal stem cells by trypsinization. Re-plate each flask so that it is split 1:6 for viral supernatant addition, as described in the sections below. It is imperative that the MSC will not become confluent during the transduction procedure, but will remain in rapid growth. Contact inhibition in adherent cells, such as MSC, increases intracellular levels of the CDK inhibitor p27, which halts cell cycle [41–43]. Target cells must traverse cell cycle to allow integration of retroviral vectors, and must be at least metabolically active for effective lentiviral vector transduction and integration [44, 45].

Figure 2.3. Over-confluent human MSC monolayer. This culture will be difficult to split evenly, due to accumulation of extracellular matrix proteins. It cannot be transduced well due to contact inhibition, which prevents or limits further division of the cells.

5. MSC TRANSDUCTION USING RETROVIRAL AND LENTIVIRAL VECTORS

For retroviral transduction, add supernatant from MoMuLV- based retroviral vectors with protamine sulfate (final concentration = 4 ug/ml) four times, over a 48 hour period. Protamine sulfate is a polycationic compound which neutralizes the negatively charged retroviral particles and cell surfaces. Add it only once every 24 hours, or it will be toxic. This should result in 20–40% of the flask being transduced, due to the rapid division of the MSC. The cells must be subconfluent when each aliquot of supernatant is added. Confluent cells (Figure 2.3) are contact inhibited and will not divide to allow retroviral vector integration.

VSV-G pseudotyped lentiviral vector supernatant can be added once or twice at an MOI of 10–100, without the need for protamine sulfate. Select the cells according to the selectable marker included in the chosen vector (if using G418 to select for the neo gene, the best concentration is 0.75 mg/ml active drug), or use as partially-transduced monolayers. Transduced MSC are excellent vehicles from which to secrete proteins, as we have described [3, 15, 46–48].

While the methodologies for transducing MSC are relatively simple, since in log phase the cells are rapidly dividing and incorporate vector very easily, in comparison to hematopoietic stem cells [18, 39, 49], several cautions do exist for transduction and for reliably assessing the success of the MSC transduction. Although lentiviral vectors

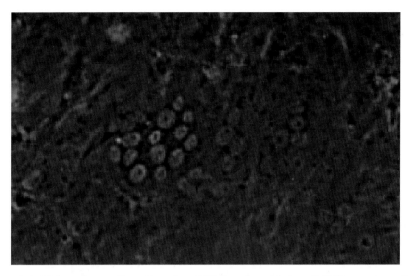

Figure 2.4. Multinucleate cells resulting from VSV-G-mediated fusion of MSC in an overconfluent culture. It is necessary to transduce subconfluent monolayers when using viral vectors.

can enter non-dividing cells, MSC monolayers should still be subconfluent prior to transduction, or the VSV-G envelope can cause cell fusion, resulting in multinucleate cells (Figure 2.4) which appear overnight in the culture.

An important caution in interpretation of transduction is that MSC can and do "share" proteins with neighboring cells, through junction formation or other as-yet-unknown mechanisms. For this reason, we have described that MSC (marrow stromal cells) must be plated at subconfluency for selective agents such as G418 to work effectively [3, 15]. This is also reflected in the fact that fluorescent markers such as eGFP can be shared between cells. Transduced cells dropped into a confluent plate of non-transduced MSC can cause a green "halo" to be seen in neighboring cells, although it is not as bright in intensity as seen in the cell that is expressing the transgene. For this reason, caution should be exercised when interpreting immunofluorescence or FACS data from partially transduced MSC cultures. If stringent parameters are set for the highly expressing cells, those that contain transgene product assimilated using the "bystander effect" will not be included. The propensity for MSC to share proteins is, however, a factor in making them excellent vehicles for delivering enzyme products or other transgenes to cells in a deficient animal or in an injured tissue.

6. STEM CELLS FOR TISSUE REPAIR: FUTURE DIRECTIONS IN THE MSC FIELD

As previously discussed in this chapter, and elsewhere in this volume, human bone marrow derived mesenchymal stem cells (BM-MSC) have the potential to form bone, cartilage, tendon, fibroblast, fat, and muscle, and may have other very exciting potentials such as contributing to the repair of damaged heart and skeletal muscle, liver, pancreas, kidney, spinal cord, and even brain. As discussed later in this volume, the *in vitro* and *in vivo* characteristics of human adipose-derived stem cells (A-MSC) have also been determined. AMSC are a novel type of cell that appear to have capacities similar to BM-MSC. Our group has learned that A-MSC home into multiple tissues in immune deficient mice, including brain (Meyerrose et al., manuscript in preparation). Therefore, if we can engineer these cells with therapeutic proteins and take advantage of their ability to home to areas of organ damage, we may be able to deliver them intravenously to promote tissue repair. The possibility of repairing tissues from easily harvested bone marrow or from unwanted fat cells holds broad appeal, and is an intriguing possibility that could have dramatic effect on health care.

Remaining questions regarding the use of MSC in tissue repair therapies are the following: (1) How are the most primitive MSC best isolated? Phenotype vs. function must be considered. (2) How are MSC recruited to the sites of tissue damage, and can this recruitment be enhanced to better accomplish cellular therapy? (3) What are the signals regulating differentiation, fusion, or recruitment of endogenous stem cells once MSC traffic into damaged tissue, and can these signals be controlled to better accomplish regenerative medicine? (4) How can the potential of MSC to promote revascularization at sites of injury be best controlled?

The remainder of this book describes different applications of MSC technology, and provides early data on the first MSC clinical trials that have been done in the United States.

REFERENCES

[1] Pittenger MF, Mackay AM, Beck SC, et al. Multilineage potential of adult human mesenchymal stem cells. *Science*. 1999;284:143–147.

[2] Dragoo JL, Lieberman JR, Lee RS, et al. Tissue-engineered bone from BMP-2-transduced stem cells derived from human fat. *Plast Reconstr Surg*. 2005;115:1665–1673.

[3] Dao MA, Pepper KA, Nolta JA. Long-term cytokine production from engineered primary human stromal cells influences human hematopoiesis in an *in vivo* xenograft model. *Stem Cells*. 1997;15:443–454.

[4] Wu GD, Nolta JA, Jin YS, et al. Migration of mesenchymal stem cells to heart allografts during chronic rejection. *Transplantation*. 2003;75:679–685.

[5] Redman SN, Oldfield SF, Archer CW. Current strategies for articular cartilage repair. *Eur Cell Mater*. 2005;9:23–32; discussion 23–32.

[6] Hui JH, Ouyang HW, Hutmacher DW, Goh JC, Lee EH. Mesenchymal stem cells in musculoskeletal tissue engineering: a review of recent advances in National University of Singapore. *Ann Acad Med Singapore*. 2005;34:206–212.

[7] Smits AM, van Vliet P, Hassink RJ, Goumans MJ, Doevendans PA. The role of stem cells in cardiac regeneration. *J Cell Mol Med*. 2005;9:25–36.

[8] Phinney DG, Isakova I. Plasticity and therapeutic potential of mesenchymal stem cells in the nervous system. *Curr Pharm Des*. 2005;11:1255–1265.

[9] Kan I, Melamed E, Offen D. Integral therapeutic potential of bone marrow mesenchymal stem cells. *Curr Drug Targets*. 2005;6:31–41.

[10] Noort WA, Kruisselbrink AB, in't Anker PS, et al. Mesenchymal stem cells promote engraftment of human umbilical cord blood-derived CD34(+) cells in NOD/SCID mice. *Exp Hematol*. 2002;30:870–878.

[11] Maitra B, Szekely E, Gjini K, et al. Human mesenchymal stem cells support unrelated donor hematopoietic stem cells and suppress T-cell activation. *Bone Marrow Transplant*. 2004;33:597–604.

[12] Lazarus HM, Koc ON, Devine SM, et al. Cotransplantation of HLA-identical sibling culture-expanded mesenchymal stem cells and hematopoietic stem cells in hematologic malignancy patients. *Biol Blood Marrow Transplant.* 2005;11:389–398.

[13] Koc ON, Gerson SL, Cooper BW, et al. Rapid hematopoietic recovery after coinfusion of autologous-blood stem cells and culture-expanded marrow mesenchymal stem cells in advanced breast cancer patients receiving high-dose chemotherapy. *J Clin Oncol.* 2000;18:307–316.

[14] Angelopoulou M, Novelli E, Grove JE, et al. Cotransplantation of human mesenchymal stem cells enhances human myelopoiesis and megakaryocytopoiesis in NOD/SCID mice. *Exp Hematol.* 2003;31:413–420.

[15] Nolta JA, Hanley MB, Kohn DB. Sustained human hematopoiesis in immunodeficient mice by cotransplantation of marrow stroma expressing human interleukin-3: analysis of gene transduction of long-lived progenitors. *Blood.* 1994;83:3041–3051.

[16] Dao MA, Hannum CH, Kohn DB, Nolta JA. FLT3 ligand preserves the ability of human CD34+ progenitors to sustain long-term hematopoiesis in immune-deficient mice after *ex vivo* retroviral-mediated transduction. *Blood.* 1997;89:446–456.

[17] Dao MA, Hashino K, Kato I, Nolta JA. Adhesion to fibronectin maintains regenerative capacity during *ex vivo* culture and transduction of human hematopoietic stem and progenitor cells. *Blood.* 1998;92:4612–4621.

[18] Dao MA, Hwa J, Nolta JA. Molecular mechanism of transforming growth factor beta-mediated cell-cycle modulation in primary human CD34(+) progenitors. *Blood.* 2002;99:499–506.

[19] Meisel R, Zibert A, Laryea M, Gobel U, Daubener W, Dilloo D. Human bone marrow stromal cells inhibit allogeneic T-cell responses by indoleamine 2,3-dioxygenase-mediated tryptophan degradation. *Blood.* 2004;103:4619–4621.

[20] Chung NG, Jeong DC, Park SJ, et al. Cotransplantation of marrow stromal cells may prevent lethal graft-versus-host disease in major histocompatibility complex mismatched murine hematopoietic stem cell transplantation. *Int J Hematol.* 2004;80:370–376.

[21] Potian JA, Aviv H, Ponzio NM, Harrison JS, Rameshwar P. Veto-like activity of mesenchymal stem cells: functional discrimination between cellular responses to alloantigens and recall antigens. *J Immunol.* 2003;171:3426–3434.

[22] Tse WT, Pendleton JD, Beyer WM, Egalka MC, Guinan EC. Suppression of allogeneic T-cell proliferation by human marrow stromal cells: implications in transplantation. *Transplantation.* 2003;75:389–397.

[23] Le Blanc K, Rasmusson I, Gotherstrom C, et al. Mesenchymal stem cells inhibit the expression of CD25 (interleukin-2 receptor) and CD38 on phytohaemagglutinin-activated lymphocytes. *Scand J Immunol.* 2004;60:307–315.

[24] Rasmusson I, Ringden O, Sundberg B, Le Blanc K. Mesenchymal stem cells inhibit lymphocyte proliferation by mitogens and alloantigens by different mechanisms. *Exp Cell Res.* 2005;305:33–41.

[25] Studeny M, Marini FC, Dembinski JL, et al. Mesenchymal stem cells: potential precursors for tumor stroma and targeted-delivery vehicles for anticancer agents. *J Natl Cancer Inst.* 2004;96:1593–1603.

[26] Nakamizo A, Marini F, Amano T, et al. Human bone marrow-derived mesenchymal stem cells in the treatment of gliomas. *Cancer Res.* 2005;65:3307–3318.

[27] Nolta JA, Dao MA, Wells S, Smogorzewska EM, Kohn DB. Transduction of pluripotent human hematopoietic stem cells demonstrated by clonal analysis after engraftment in immune-deficient mice. *Proc Natl Acad Sci U S A.* 1996;93:2414–2419.

[28] Dao MA, Shah AJ, Crooks GM, Nolta JA. Engraftment and retroviral marking of CD34+ and CD34+CD38- human hematopoietic progenitors assessed in immune-deficient mice. *Blood.* 1998;91:1243–1255.

[29] Dao MA, Yu XJ, Nolta JA. Clonal diversity of primitive human hematopoietic progenitors following retroviral marking and long-term engraftment in immune-deficient mice. *Exp Hematol.* 1997;25:1357–1366.

[30] Boiret N, Rapatel C, Veyrat-Masson R, et al. Characterization of nonexpanded mesenchymal progenitor cells from normal adult human bone marrow. *Exp Hematol.* 2005;33:219–225.

[31] Ahrens N, Tormin A, Paulus M, et al. Mesenchymal stem cell content of human vertebral bone marrow. *Transplantation.* 2004;78:925–929.

[32] Zhang Y, Li CD, Jiang XX, Li HL, Tang PH, Mao N. Comparison of mesenchymal stem cells from human placenta and bone marrow. *Chin Med J (Engl).* 2004;117:882–887.

[33] Igura K, Zhang X, Takahashi K, Mitsuru A, Yamaguchi S, Takashi TA. Isolation and characterization of mesenchymal progenitor cells from chorionic villi of human placenta. *Cytotherapy.* 2004;6:543–553.

[34] O'Connell PJ, McKenzie A, Fisicaro N, Rockman SP, Pearse MJ, d'Apice AJ. Endoglin: a 180-kD endothelial cell and macrophage restricted differentiation molecule. *Clin Exp Immunol.* 1992;90:154–159.

58 T. MEYERROSE ET AL.

[35] Lastres P, Bellon T, Cabanas C, et al. Regulated expression on human macrophages of endoglin, an Arg-Gly-Asp-containing surface antigen. *Eur J Immunol.* 1992;22:393–397.

[36] Roy-Chaudhury P, Simpson JG, Power DA. Endoglin, a transforming growth factor-beta-binding protein, is upregulated in chronic progressive renal disease. *Exp Nephrol.* 1997;5:55–60.

[37] Chen CZ, Li L, Li M, Lodish HF. The endoglin(positive) sca-1(positive) rhodamine(low) phenotype defines a near-homogeneous population of long-term repopulating hematopoietic stem cells. *Immunity.* 2003;19:525–533.

[38] Chen CZ, Li M, de Graaf D, et al. Identification of endoglin as a functional marker that defines long-term repopulating hematopoietic stem cells. *Proc Natl Acad Sci U S A.* 2002;99:15468–15473.

[39] Dao MA, Taylor N, Nolta JA. Reduction in levels of the cyclin-dependent kinase inhibitor p27(kip-1) coupled with transforming growth factor beta neutralization induces cell-cycle entry and increases retroviral transduction of primitive human hematopoietic cells. *Proc Natl Acad Sci U S A.* 1998;95:13006–13011.

[40] Devine SM, Flomenberg N, Vesole DH, et al. Rapid mobilization of CD34+ cells following administration of the CXCR4 antagonist AMD3100 to patients with multiple myeloma and non-Hodgkin's lymphoma. *J Clin Oncol.* 2004;22:1095–1102.

[41] Polyak K, Kato JY, Solomon MJ, et al. p27Kip1, a cyclin-Cdk inhibitor, links transforming growth factor-beta and contact inhibition to cell cycle arrest. *Genes Dev.* 1994;8:9–22.

[42] Kato A, Takahashi H, Takahashi Y, Matsushime H. Contact inhibition-induced inactivation of the cyclin D-dependent kinase in rat fibroblast cell line, 3Y1. *Leukemia.* 1997;11(suppl 3):361–362.

[43] Fuse T, Tanikawa M, Nakanishi M, et al. p27Kip1 expression by contact inhibition as a prognostic index of human glioma. *J Neurochem.* 2000;74:1393–1399.

[44] Miller DG, Adam MA, Miller AD. Gene transfer by retrovirus vectors occurs only in cells that are actively replicating at the time of infection [published erratum appears in *Mol Cell Biol.* 1992 ;12(1):433]. *Mol Cell Biol.* 1990;10:4239–4242.

[45] Sutton RE, Reitsma MJ, Uchida N, Brown PO. Transduction of human progenitor hematopoietic stem cells by human immunodeficiency virus type 1-based vectors is cell cycle dependent. *J Virol.* 1999;73:3649–3660.

[46] Dao MA, Nolta JA. Inclusion of IL-3 during retrovirally-mediated transduction on stromal support does not increase the extent of gene transfer into long-term engrafting human hematopoietic progenitors. *Cytokines Cell Mol Ther.* 1997;3:81–89.

[47] Tsark EC, Dao MA, Wang X, Weinberg K, Nolta JA. IL-7 enhances the responsiveness of human T cells that develop in the bone marrow of athymic mice. *J Immunol.* 2001;166:170–181.

[48] Wang X, Ge S, McNamara G, Hao QL, Crooks GM, Nolta JA. Albumin expressing hepatocyte-like cells develop in the livers of immune-deficient mice transmitted with highly purified human hematopoietic stem cells. *Blood.* 2003.

[49] Dao M, Nolta J. Molecular control of cell cycle progression in primary human hematopoietic stem cells: methods to increase levels of retroviral-mediated transduction. *Leukemia.* 1999;13:1473–1480.

CHAPTER 3

GENE EXPRESSION PROFILES OF MESENCHYMAL STEM CELLS

D. G. PHINNEY

*Centre for Gene Therapy, Tulane University Health Sciences Center,
New Orleans, LA.*

1. INTRODUCTION

Nearly 40 years ago, Friedenstein and co-workers first demonstrated that bone marrow is inherently osteogenic, capable of generating a heterotopic ossicle *in vivo* that supports host cell hematopoiesis and is self-maintaining and self-renewing [reviewed in 1]. Friedenstein hypothesized that this property of marrow was attributed to the existence of an osteogenic stem cell and later validated this hypothesis by showing that fibroblastoid cells enriched from bone marrow via their attachment to plastic retained the capacity to form bone and cartilage when implanted *in vivo* [2]. Subsequent to these seminal discoveries, various laboratories demonstrated that these adherent marrow fibroblasts, referred to as stromal cells, were capable of differentiating into adipocytes, chondrocytes, and osteoblasts using *in vitro* assays [3–10]. Collectively, these studies indicated that Friedenstein osteogenic stem cells were in actuality a mixture of multi-potent mesenchymal progenitors and/or mesenchymal stem cells (MSCs). Ensuing work demonstrating that rodent [11] and human [12] stromal cell populations derived from single cells exhibited multilineage differentiation *in vitro* supported the latter contention, and the term MSC was adopted in the literature to describe this unique cell population.

As the aforementioned studies indicate, MSCs are typically classified according to their functional characteristics. Notably, a specific molecular phenotype has not been ascribed to these stem cells. The latter is partly attributed to the lack of *in vivo* assays for evaluating the repopulating ability and differentiation potential of cells purified using prospective stem cell markers. This is particularly frustrating in that the ability of MSCs to undergo multi-lineage differentiation *in vitro* does not necessary reflect their degree of pluripotency *in vivo* [13, 14]. Efforts to ascribe a phenotype to MSCs have also been confounded by the fact that most preparations are a heterogeneous mixture of uni-, bi-, and multi-potent cells [13, 15, 16]. Moreover, cultured MSCs express various

J.A. Nolta (ed.), Genetic Engineering of Mesenchymal Stem Cells, 59–80.

cell-lineage specific antigens *in vitro* that vary between different preparations and as a function of time in culture, but do not correlate with changes in differentiation potential [14, 17]. Accordingly, identifying a molecular fingerprint for MSCs has been likened to "shooting at a moving target" [14]. This review provides a summary of those genes reported to be expressed in MSCs, and discusses how they relate to the biology and function of the bone marrow stroma.

2. GENE EXPRESSION PROFILES OF MARROW STROMAL CELLS

2.1. Long term bone marrow cultures (LTBMCs)

Friedenstein demonstration that marrow stromal cells produced an environment *in vivo* conducive for hematopoiesis illustrated their potential usefulness for developing culture systems that recapitulate this process *in vitro*. In the 1970s, Dexter and co-workers [18] showed that long term bone marrow cultures (LTBMCs), which were prepared by charging an established monolayer culture of stromal cells with fresh bone marrow, supported production of CFU-S, GM-CFC, granulocytes, and monocytes. These cultures were later modified to also support production of megakaryocytic and erythroid progenitors [19, 20] as well as B cell progenitors [21]. Subsequently, human LTBMCs were established that could sustain hematopoiesis for up to 20 weeks [22]. The development of murine and human LTBMCs provided a unique opportunity to dissect the cell-type specific interactions and soluble factors that regulate aspects of hematopoiesis. Analysis of these cultures provided a wealth of information regarding the phenotype and function of marrow stromal cells.

A widely recognized property of LTBMCs is that the most primitive hematopoietic progenitors are tightly attached to the stromal cell layer, and only with increasing maturity are they found to migrate into the surrounding media [23–26]. Accordingly, stromal cells have been shown to secrete an array of matrix molecules, including collagens, fibronectin, laminin, vitronectin, thrombospondin, haemonectin, thrombopoietin, tenascin as well as other proteoglycans and glycosaminoglycans that function as binding sites for hematopoietic progenitors and the mitogens and cytokines that stimulate their growth and maturation [27–32]. Additionally, specific ligand/receptor interactions between stromal cells and hematopoietic cells were identified that were deemed critical for sustained hematopoiesis *in vitro*. For example, MSCs express VCAM-1 [33] and produce fibronectin and thrombospondin, which are ligands for the integrin heterodimer $\alpha 4\beta 1$ (CD49d/CD29) known as Very Late Activation antigen-4 (VLA-4) (34). This integrin receptor is expressed by virtually all CD34+ cells from bone marrow, cord or peripheral blood [35–37]. Various studies have shown that CD34+ progenitors as well as BFU-E, GM-CFCs, and B cell precursors exhibit VLA-4 dependent binding to the stromal cell layer in LTBMCs [38–41]. Moreover, anti-VLA-4 monoclonal antibodies block lymphopoiesis and myelopoiesis and inhibit formation of erythroblastic islands when added to murine LTBMCs [42–44]. The importance of this interaction has also been demonstrated *in vivo*, as administration of anti-VLA-4 monoclonal antibodies to non-human primates leads to an 8 to 200-fold increase in the number of mobilized hematopoietic progenitors in blood [45]. Similar adhesion blocking experiments have shown that expression of CD44 and ICAM-1 by stromal cells is also necessary for sustained hematopoiesis in LTBMCs [41, 46]. Other stromal cell proteins that mediate adhesion or regulate survival and maturation of hematopoietic progenitors in LTBMCs include Flt-3 ligand [47], hepatocyte growth factor [48], Jagged1 [49] neuropilin-1 [50], CD164 [51], CD28 [52], CD49d and CD90 [53] and the laminin gamma2 chain [54].

Marrow stromal cells have also been shown to constitutively express an array of cytokines, including stem cell factor, GM-CSF, G-CSF, M-CSF, KL, LIF, IL-1β, IL-3, IL-6, IL-7, IL-8, IL-11, IGF, and TGFβ that support the growth and maturation of hematopoietic cells [55, 56]. Expression levels of these cytokines may also be altered in response to external stimuli. For example, phorbol myristate acetate and TNF-α induce stromal cells to produce activin A, which stimulates proliferation of hematopoietic stem cells [57]. Exposure to IL-7 induces IL-6 expression [58] and IL-1β, IL-6, and lipopolysaccharides induce increased levels of GM-CSF and G-CSF in stromal cells [59]. Alternatively, treatment of stromal cells with interferon α has been shown to down regulate expression of these colony stimulating factors. This responsiveness in the paracrine production of cytokines by stromal cells is thought to provide a means to alter hematopoiesis in response to stress, infection, injury, and other insults.

2.2. Molecular characterization of stromal cells

In addition to factors that regulate hematopoiesis, stromal cells have been shown to express the LDL receptor and alkaline phosphatase [60], smooth muscle actin [61], type IV collagen and laminin [62], factor VIII [63], and MUC18 [64]. Consequently, stromal cells have been described as marrow myoid cells, vascular smooth muscle cells, or endothelial-like cells, spurring a debate about their nature and ontogeny. Stromal cells also reportedly express Fas [65], various integrin proteins [66], Nemo-like kinase [67], Leptin [68], various insulin-like growth factor binding proteins [69], the receptor tyrosine kinases PDGFR-β, EGFR, FGFR1, and Axl [70] as well as possess low- and high-voltage-activated Ca2+ currents [71]. In contrast, the cells were shown to be devoid of most markers common to hematopoietic cell types, including CD2, CD3, CD4, CD8, Mac-1/CD11b, CD14, CD15, CD19, CD20, B220, CD45, Thy-1, and myeloperoxidase [56]. However, conflicting reports exists regarding whether or not stromal cells express the glycoprotein CD34 [72–74].

3. GENE EXPRESSION PROFILES OF MESENCHYMAL STEM CELLS

3.1. Molecular characterization of MSCs

The reclassification of stromal cells as MSCs spurred renewed interest in the biology of these plastic adherent populations. Not surprisingly, the growing list of markers reported to be expressed by MSCs reiterate many past characterizations of marrow stromal cells. For example, the antibody Stro-1 was originally shown to be expressed by erythroid progenitors as well as stromal elements with the capacity to support hematopoiesis [75]. Subsequently, the STRO-1 fraction of bone marrow was shown to contain pre-osteoblastic cells, while those lacking STRO-1 were deemed characteristic of fully differentiated osteoblasts [76]. Most recently, marrow fibroblastoid cells isolated using magnetic immuno-beads linked to STRO-1 were shown to be capable of multi-lineage differentiation [77] indicating that this antibody enriches for a cell population with the functional characteristics of MSCs. Clonegenic adherent marrow cells with the capacity for adipogenic and osteogenic differentiation have also been isolated by immuno-selection using antibodies against nerve growth factor receptor [78]. Similarly, this receptor was previously shown to be expressed by a specialized type of stromal cell in bone marrow that is distinguished morphologically by its dendrite-like processes [79]. Other antibodies reported to bind MSCs include SB-10, which recognizes an epitope of ALCAM (CD166) [80]; SH2, which recognizes an epitope on Endoglin (CD105) [81]; and SH-3/SH-4, which recognize distinct epitopes on CD73 [82].

Pittenger et al. [83] recently described methods to isolate and culture expand human MSC populations that uniformly express CD29, CD44, CD73 (SH-3), and CD105 (SH-2) but lack expression of CD14, CD34, and CD45. Analysis by flow cytometry identified over 50 growth factors, integrins, cytokine receptors and matrix molecules expressed by these cells, many of which were previously identified in stromal cells. Although this study represents one of the most complete characterizations of human MSCs, no surface marker was identified that alone was sufficient to distinguish the stem cells.

Human MSCs with the aforementioned phenotype have also been shown to constitutively express a diverse array of cytokines and support hematopoiesis *in vitro* [84, 85], apparent attributes of their "stromal cell" heritage. In contrast, these and other human MSC preparations have been shown to lack expression of telomerase [86, 87], an enzyme known to be expressed in stem cells. When expressed ectopically, telomerase enabled MSCs to undergo more than 260 population doublings *in vitro* without altering their karyotype [89] and yield greater bone production *in vivo*, presumably due to expansion of the osteoprogenitor pool [90]. A recent study has shown that FGF2 treatment induces a transient increase in the telomere length, extends the life span, and prolongs the differentiation potential of *ex vivo* expanded human MSCs [88]. Therefore, FGF2 may specifically promote expansion of a rare, telomerase expression stem cell population in these cultures. The latter is consistent with studies in our laboratory showing that FGF2 induces proliferation and reversibly inhibits the differentiation of murine MSCs [74]. In an unrelated study, Jia et al. [91] generated 4258 ESTs by single-pass sequencing from a human MSC cDNA library, which included 1860 unique sequences. The 30 most abundant expressed genes included matrix proteins characteristic of connective tissues, such as fibronectin, collagens type I, III, and V, osteonectin, decorin, and vimentin. The library also contained 60 ESTs representing novel genes not found in other libraries but the identity of these unique transcripts has not been reported.

3.2. Gene expression in MSCs following cellular differentiation

A significant body of literature has been devoted to identifying transcripts up regulated in MSCs in response to stimuli that induce cellular differentiation [92]. For example, dexamethasone has been shown by real-time PCR [93] to reliably induce expression of BMP-2, osteopontin, bone sialoprotein, and Cbfa1 and by differential display [94] to alter the expression of osteonectin, collagen type III, fibronectin, SM22, calphobindin

II, cytosolic thyroid hormone binding protein, and a TGF-β induced transcript in human MSCs. Comparative studies employing DNA micro arrays revealed 55 transcripts that were up regulated and 82 transcripts that were down regulated in human MSCs by dexamethasone [95]. The former included genes already known to be involved in mineral metabolism, shown previously to be induced by dexamethasone, or previously associated with osteogenic differentiation. Transcripts that were down regulated included tropomyosin 2, myosin regulatory light chain 2, other muscle-related genes, matrix metalloproteinase 14 and inhibitor of differentiation-4 (Id4). Interestingly, expression of Id4, together with Id1-3 were shown to be down regulated in an immortalized human marrow stromal cell line following exposure to BMP-2, suggesting that these dominant negative helix-loop-helix proteins may block MSC differentiation in the absence of specific stimuli [96]. Treatment of MSCs with BMP-2 was also shown to induce expression of *dlx-2, hes-1, stat1, junB, sox4, areb6,* and *cbfa1* in these studies. DNA micro arrays have also been used to delineate the molecular events associated with chondrogenic differentiation of human MSCs [97].

Specific signalling pathways that regulate the growth and differentiation of human MSCs have also been characterized. For example, Jaiswal et al. [98] showed that activation or inhibition of extracellular regulated kinase (ERK) directs differentiation of MSCs to the osteogenic or adipogenic lineage, respectively. Human MSCs have also been shown to express high levels of Dickkopf 1 (DKK1) during the transition form lag phase to exponential growth and low levels of DKK1 during cell cycle arrest following serum deprivation [99]. Therefore, DKK1 appears to stimulate entry of MSCs into the cell cycle by inhibiting the canonical Wnt signalling pathway and decreasing cellular concentrations of beta-catenin.

3.3. Gene expression and MSC plasticity

Studies examining the potential of MSCs to differentiate into non-mesodermal lineages have shown that the cells express epithelial and neural specific transcripts. For example, human MSCs were shown to up regulate expression of CC26, E-cadherin, β-catenin, and cytokeratins 17, 18, and 19 following co-culture *in vitro* with small airway epithelial cells, but were also shown to constitutively express cytokeratins 8, 10, and 18 [100]. Studies evaluating the potential of MSCs to differentiate into neurons *in vitro* revealed the cells constitutively express low levels of neuron-specific nuclear protein (NeuN) and GFAP [101] as well as neuron specific enolase [102]. More recent reports have confirmed and extended these observations, showing that MSCs express transcripts corresponding to the NMDA glutamate binding subunit, syntaxin, GFAP, neuroD, nestin and neurofilament proteins [103–106]. Expression of these transcripts by MSCs is thought to presage their ability to differentiate into ectodermal cell lineages *in vitro*.

3.4. MSCs and the mesengenic process

One obvious characteristic made apparent by the aforementioned studies is that MSCs simultaneously function as hematopoiesis-supporting stroma and multi-potent stem cells. This duality of function contradicts the typical hierarchical process of stem cell differentiation typified by the mesengenic process proposed by Caplan [107]. In this scheme MSCs sits atop a hierarchy of progressively more determined progenitors that that yield cell types of specific function, including hematopoiesis-supporting stromal cells. Based on this paradigm, one would expect that MSCs and stromal cells would be readily discernable based on phenotype and function. Their lack of distinction is exemplified by a recent study, which revealed only minor differences in cytokine expression and capacity to support hematopoiesis between these populations [108]. This disparity between the functional characteristics of MSCs implied by the mesengenic process and observed experimentally would indicate that their biology does not conform to established stem cell paradigms. This notion is supported by the fact that determined mesenchymal cell types, such as chondrocytes and adipocytes, have been shown to undergo trans-differentiation, or switch from one cell fate to another, in response to external cues [14]. These unorthodox properties continue to confound efforts to define the nature of the MSC. Moreover, trying to develop a portrait of MSCs based on gene expression profiles is precarious, as most studies employ differing methods to isolate and culture expand the cells and survey only a small fraction of the total transcript pool.

4. THE MSC TRANSCRIPTOME

4.1. Serial analysis of gene expression (SAGE)

To better characterize the molecular phenotype of MSCs, we catalogued their repertoire of expressed transcripts using a high throughput transcript profiling methodology known as serial analysis of gene expression (SAGE) [109]. This technique employs a series of enzymatic reactions to convert expressed mRNAs into 9–13 bp cDNA fragments, or tags. Because a 10 bp tag can theoretically distinguish 4^{10} (1, 048, 576) transcripts, each tag essentially represents a unique identifier for its cognate mRNA. Once generated, the tags are concatenated into long strings, cloned, and sequenced thereby accelerating the process of gene identification. This technique is advantageous because it records all mRNA sequences directly, including those not yet cloned and deposited in public databases. The tag frequency in a given SAGE library has also been shown to approximate the abundance of its cognate mRNA in the specimen analyzed [110]. Therefore, SAGE is both comprehensive and quantitative.

4.2. Transcriptome of a single cell-derived colony of human MSCs

Initially, we used SAGE to analyze the molecular phenotype of a clonally-derived population of human MSCs [17]. In these studies we cultured plastic adherent bone marrow cells under condition that select for high colony forming efficiency and multipotency [111] and isolated a single cell-derived clone of approximately 10,000 cells from these cultures. We then used a modified MicroSAGE protocol [112] to analyze the transcriptome of the colony. In total, we sequenced and catalogued 17, 767 tags of which 6, 977 were unique. The 50 most abundant expressed transcripts included proteins characteristic of connective tissues such as vimentin, fibronectin, MMP2, collagens type I, III, and VI as well as the transcriptional regulator Cbfa1, consistent with most previous findings. Further interrogation of the database revealed that the single cell-derived colony also expressed mRNAs characteristic of determined mesenchymal cell lineages. For example, 1.6% of the tags corresponded to skeletal-specific transcripts and 1.36% to muscle-specific transcripts. In the former case we identified unique SAGE tags representing 30 skeletal-specific genes including those encoding osteonectin, osteopontin, alkaline phosphatase, cartilage matrix protein, collagen type IX, cadherin-11, vitamin D receptor, and the BMP receptor type II. Similarly, we identified unique SAGE tags corresponding to 55 genes encoding muscle specific proteins, including the bradykinin receptor, calponin, fukutin, nebulin, troponin C, tropomodulin, tropomyosin 1–4, myocyte enhancer factor 2A and 2C, muscle specific isoforms of glycogen synthase 1, phosphofructokinase, and pyruvate kinase as well as skeletal, cardiac, and smooth muscle isoforms of myosin. We also identified unique SAGE tags corresponding to transcripts expressed by adipocytes and hematopoiesis-supporting stroma. Therefore, the transcriptome of the single cell-derived colony reflected the developmental potential of MSCs as defined by functional assays.

Our analysis was also the first to report that MSCs express mRNAs characteristic of endothelial, epithelial, and neural cell lineages. These included the endothelial specific proteins MUC18, VEGFR, endoglin, podocalyxin-like protein, endothelial differentiation related factor 1 and endothelial cell growth factor 1. Epithelial specific proteins included keratins 8, 10 and 18, epican, epithelial membrane protein 3, and several gap junction proteins. The colony also expressed transcripts encoding neurotransmitter receptors, factors involved in synaptic transmission, such as intersectin, synapsin III, and catechol-O-methyltransferase as well as other proteins common in neural tissue, including N-cadherin, peripheral myelin protein 22, prosaposin, GFAP, and glia maturation factor β. Since this analysis was done on a single cell-derived clone, it ruled out the possibility that these transcripts were contributed from contaminating cell types in the preparation.

To more extensively characterize the MSC transcriptome, we developed a software module termed Tag Macro Suite (TMS) v1.4 consisting of seven separate macros written in Visual Basic that runs in Microsoft®Excel. The software, which will be described in detail elsewhere, manipulates SAGE data in a manner that facilitates interrogation using multiple formats, such as searching for a specific keyword or genes grouped according to function. The software can also be used to simultaneously compare multiple SAGE databases. Using TMS v1.4 we determine the percentage of SAGE tags corresponding to genes that have specific molecular functions as defined by the Celera®Discovery System Panther classifications (Table 3.1). The most striking aspect of the MSC transcriptome revealed by this analysis was the large percentage of tags, 13.6% denoted as "No Reliable Match," that fail to match any known genes in the public database. Additionally, 5% of tags correspond to expressed sequence tags (ESTs). Therefore, a significant percentage of the MSC colony's transcriptome corresponds to novel gene sequences. Based on this classification scheme, we identified unique

Table 3.1. Percentage of SAGE tags that correspond to transcripts categorized according to molecular function and the total number of unique genes identified for each category.

	Colony		Population	
Molecular function	Total tags (%)	Genes	Total tags (%)	Genes
Cytoskeletal Proteins	3.99	64	6.03	126
ESTs	4.92	ND	3.07	ND
Extracellular Matrix Proteins	3.91	35	3.5	52
Kinases	2.54	96	2.76	168
No Reliable Match	13.64	ND	11.01	ND
Phosphatases	1.23	33	1.7	68
Proteases	3.31	60	2.63	95
Receptors	2.33	81	2.0	117
Regulatory Molecules	5.00	127	5.73	205
Signalling Molecules	3.99	55	4.54	108
Transcription Factors	4.74	141	4.50	280

SAGE tags corresponding to 64 cytoskeletal proteins, 35 extracellular matrix proteins, 96 kinases, 33 phosphatases, 60 proteases, 81 receptors, 127 regulatory molecules, 55 signalling molecules, and 141 transcription factors, examples of which are listed in Table 3.2.

To determine a molecular fingerprint for the MSC colony, we compared its transcriptome to that of 24 different human cell types and tissues for which SAGE databases are available in the public domain (Table 3.3). The 25 SAGE databases, including the MSC colony, contained a total of 39,774 unique SAGE tags that occurred with a frequency of two or greater. A total of 2320 tags were found exclusively in the MSC colony database, and they represented 1960 distinct transcripts. Strikingly, 917 (45%) of these tags corresponded to unknown genes, 206 (10%) corresponded to ESTs, and 101 (5%) matched hypothetical proteins. Therefore, approximately 61% of all SAGE tags unique to the MSC colony, as defined by the criteria outlined above, corresponded to unknown genes. Some of the remaining SAGE tags matched transcripts encoding proteins found principally in skeletal tissues, such as dermatopontin and integrin $\alpha 11$. Tags corresponding to the proteosome protein sequestosome 1 were also found. Interestingly, mutations in this protein have been shown to cause Paget disease, a bone disorder characterized by disorganized increases in bone turnover [113]. Other tags matched the signalling proteins secreted frizzled related protein-2, frizzled homolog 4, and smad7 as well as the transcription factors *sox18*, *cbfa1*, short stature homeobox 2 (*sshox2*) and *twist*. Interestingly, *sshox2* has been shown to play a role in craniofacial, heart, and limb development [114]. Similarly, *twist* plays an important role in the specification of mesoderm during development [115], is expressed in somites, head mesenchyme, and limb buds during mouse embryogenesis [116], and has been shown to regulate osteogenic differentiation [117, 118]. Other transcripts identified by this analysis included syncoilin, an intermediate filament protein localized at neuromuscular junctions [119], an uncharacterized hematopoietic stem cell

Table 3.2. Top five most abundant transcripts corresponding to unique SAGE tags classified according to molecular function.

Molecular function	MSC colony	MSC population
Cytoskeletal Proteins	vimentin, β-actin, thymosin β-10, membrane protein 3, zyxin	vimentin, β-actin, thymosin β-10, transgelin, transgelin 2
Extracellular Matrix Proteins	ColVIα3, MMP2, fibronectin, decorin, ColIα1	fibronectin, ColIα1, MMP2, ColVIα2, ColVIα3
Kinases	slit homolog 2, guanylate kinase 1, Integrin-linked kinase, MAPKK2, IL-1R associated kinase	integrin-linked kinase, pyruvate kinase (muscle), phosphofructokinase, FGFR-like 1, MAPKK2
Phosphatases	dual specificity phosphatase 1, phosphatidic acid phosphatase 2B, PTP IVA, dual specificity phosphatase 14, phosphatidic acid phosphatase 2A	PTP non-receptor 12, dual specificity phosphatase 1 and 14, histidine triad nucleotide binding protein 1, cdk inhibitor 3
Proteases	MMP2, plasminogen activator inhibitor type I, cathepsin K, MMP14, dipeptidylpeptidase IV(CD26)	MMP2, plasminogen activator inhibitor type I, proteosome subunit β7 and β3, MMP14
Receptors	alpha-2 macroglobulin receptor, protease nexin-II, stromal cell derived factor receptor 1, osteoprotegerin, C-type lectin	amyloid beta precursor protein, FGFR-like 1, stromal cell derived factor receptor 1, Axl, signal recognition particle receptor
Regulatory Molecules	plasminogen activator inhibitor type I, follistatin-like 1, G protein β polypeptide 2-like 1, pigment epithelium derived factor, TIMP 1	IGFBP7, ras homolog, cyclin D1, plasminogen activator inhibitor type I, follistatin-like 1, cell division cycle 42
Signalling Molecules	galectin 1, gremlin, cyr61, thy-1, follistatin-like 1	galectin 1, IGFBP7, follistatin-like 1, cyr61, TGF-β binding protein 2
Transcription Factors	v-jun sarcoma virus 17 homology, SRY-box 4, NF-κB inhibitor epsilon, paired mesoderm homeo box 1, ATF 4	elongation factor A-like 1, NF-κB inhibitor epsilon, basic transcription factor 3, enhancer of rudimentary, C-terminal binding protein 1

Table 3.3. Human SAGE databases used for comparative analysis.

Database designation	Cell type or tissue	Database designation	Cell type or tissue
GSM1	Primary foreskin fibroblasts	GSM728	Normal colonic epithelium
GSM573	Peripheral retina	GSM730	Astrocytes
GSM668	Uninduced 293 cells	GSM738	Mesothelial cells
GSM676	95% white matter from normal brain	GSM760	Normal luminar mammary epithelium
GSM677	Normal luminar mammary epithelium	GSM762	Normal lung tissue
GSM706	Vascular endothelial cells	GSM785	Normal liver
GSM708	Normal kidney tissue	GSM786	Normal cortex
GSM709	Leucocytes	GSM824	Vastus lateralis muscle biopsy
GSM711	Post-crisis skin fibroblasts	GSM1121	Epidermal keratinocytes
GSM716	Normal pancreas duct epithelial cells	GSM1123	Skin epidermis
GSM719	Ovary surface epithelium	GSM1499	Normal heart
GSM722	Ovarian surface epithelium cell line	GSM2386	Normal spinal cord

protein, craniofacial development protein 1, and a homeobox protein expressed in ES cells.

4.3. Transcriptome of a non-clonal human MSC population

We also analyzed the transcriptome of a non-clonal population of human MSCs derived from the same donor sample as the single cell-derived colony. Prior to analysis the MSC population was expanded *in vitro* for several more passages as compared to single cell-derived colony. We sequenced and catalogued 69,937 SAGE tags of which 17, 982 were unique. The transcriptome of the MSC population was remarkably similar in composition as compared to that of the single cell-derived colony (Table 3.1) and contained a high percentage of tags corresponded to unknown genes (11.01%) or ESTs (3.07%). Interrogation of the database revealed unique tags corresponding to 126 cytoskeletal proteins, 52 extracellular matrix proteins, 168 kinases, 68 phosphatases, 95 proteases, 117 receptors, 205 regulatory molecules, 108 signalling molecules, and 280 transcription factors (Table 3.2). While the colony and population appeared to express many common transcripts, some differences were notable. For example, the colony's transcriptome contained a large number of tags matching the collagen binding protein decorin and the decoy receptor osteoprotegrin. In contrast, the MSC population appeared to express high levels of the actin cross linking proteins transgelin 1 and 2 and the Axl receptor tyrosine kinase, which was previously detected in stromal cells [70]. Moreover, the colony was also distinguished by a greater number of tags corresponding to *gremlin*, a BMP antagonist expressed during limb morphogenesis [120, 121].

Further interrogation revealed that 4.1% and 2.6% of all tags matched transcripts encoding skeletal and muscle specific proteins, respectively. Therefore, the population appeared to express a higher percentage of transcripts corresponding to determined mesenchymal cell lineages as compared to the colony. However, we did not detect a significant difference in the percentage of SAGE tags corresponding to cytokines, cytokine receptors, and proteins involved in hematopoiesis between the two transcriptomes. We also identified SAGE tags matching transcripts encoding neurotrophic factors as well as proteins that mediate axons guidance, promote neuritogenesis and angiogenesis. Examples of these include brain derived neurotrophic factor, nerve growth factor, glia maturation factor β, pleiotrophin, prosaposin, pigment epithelial derived factor, semaphorins, VEGF, VEGF-B, VEGF-C, angiopoietin 1 and 2, angiomotin, and endothelial cell growth factor. The transcriptome of the population was also characterized by a high percentage of tags encoding factors involved in cell communication, motility, and neuronal activity.

A comparative analyses done as described above revealed a total of 625 tags representing 442 distinct genes unique to the MSC population. Approximately 45% of these tags corresponded to transcripts of an indeterminate nature, including 152 unknown genes, 28 ESTs and 27 hypothetical proteins. Therefore, the total number of unique tags and the percentage matching unknown genes was lower in the transcriptome of the population as compared to the colony. Tags that were unique to the population matched transcripts encoding ColVIα2, ColVα1, Cbfa1, stromal cell derived factor 1, biglycan, cytoskeleton associated protein 2, osteonidogen, endothelial cell growth factor 1, prosaposin, and axotrophin.

4.4. Colony vs. population

To further highlight differences between the MSC colony and population we compared their transcriptomes directly. Collectively, the two transcriptomes contained 21,521 unique SAGE tags of which 3351 were common to the two populations. These represented 2933 unique transcripts. Approximately 23% of the tags were of an indeterminate nature. Moreover, only 18 out of the top 50 expressed tags were common to the two libraries, indicating substantial differences in the expression levels of many genes. A partial list of transcripts whose corresponding tag frequencies varied between the two populations is given in Table 3.4. Transcripts represented by a higher frequency of SAGE tags in the colony included cathepsin K, decorin, the LDL receptor, CD109, several homeotic genes and several transcripts induced during mesoderm development. Alternatively, tags occurring at a higher frequency in the population matched transcripts encoding IGFBP 6 and 7, Axl tyrosine kinase receptor, TGF-α, biglycan, elastin, osteonectin, and ColIVα1. In addition, comparative studies identified 45 tags that were common to both MSC populations but not found in the 24 other human SAGE libraries (Table 3.3). A total of seven of these tags mapped to unknown genes and the remaining ones matched 25 distinct transcripts, including those encoding the proteins Pax5, Twist, Gremlin, Cbfa1, stromal cell derived factor 1, and integrin α11.

Table 3.4. Fold-difference in tag abundance of select transcripts common to the transcriptomes of the MSC colony and population.

Transcript identity	Fold-difference (colony/population)
Cathepsin K	126
Decorin	61
No Reliable Match	52
No Reliable Match	48
No Reliable Match, paired mesoderm homeobox 1	34.9
No Reliable Match [2]	~26
JunD	22
No Reliable Match [3], Oncostatin M receptor	17.5
Trophoblast glycoprotein	13
No Reliable Match [7], Dickkopf homolog 3, Hox B7, LDL receaptor, MAD7, mesoderm development candidate 2, mesoderm induction early response 1, sprouty 2 homolog	8.7
No Reliable Match [2], Twist, ColXVIα2, PDGFR-β	6.5
Paired related homeobox protein	5.4
CD109, frizzled homolog 4	4.3
No Reliable Match [3]	−3.2
TGF-α	−3.7
No Reliable Match [2], Laminin receptor 1, Axl	−4.0
Transgelin 2, IGFBP6, elastin	−4.8
Biglycan	−5.5
Transgelin, cyclin D1, osteonectin	−6.5
No Reliable Match [3], ColIVα1, lysyl oxidase-like 2	−15
IGFBP7	−23.6

5. DISCUSSION

5.1. The MSC colony transcriptome

Our SAGE studies revealed that the vast majority of transcripts expressed in a single cell-derived colony of MSCs correspond to unknown genes or proteins of indeterminate function. This exemplifies our lack of knowledge regarding the molecular pathways regulating MSC biology and may reflect the difficulty in ascribing a specific phenotype to these stem cells. We also catalogued SAGE tags corresponding to skeletal, muscle, adipose, and stromal specific transcripts expressed in the colony. Since these cell lineages are those specifically derived from MSCs, this finding may indicate that the MSC exits in a "ground state" as described for hematopoietic stem cells [122]. According to this model, stem cells are primed with respect to mRNA transcription, expressing low levels of transcripts reflecting the function of various differentiated cell lineages. In response to a specific inducing stimulus the cells up regulate expression of factors required for the function of a specific cell type and down regulate those characteristic of alternative cell fates. Hence, MSCs may simultaneously exhibit characteristics of bone, fat, cartilage, and muscle until external signals direct their differentiation into one lineage. Alternatively, in the absence of factors that promote stem cell self-renewal, cells of the colony may undergo lineage commitment during expansion *ex vivo*. The stochastic nature of this process would produce progenitors that express transcripts characteristic of their different cell fates. Both of these scenarios imply that the clone initiating cell used in our analysis was a stem cell or early progenitor, thereby underscoring the value of our SAGE database. Unfortunately, the only way to discriminate between these two processes is to profile the transcriptome of individual cells, which is technically challenging.

The transcriptome may also provide clues to the ontogeny of MSCs. For example, our studies show that a single cell-derived colony of MSCs is characterized by expression of mesenchymal, endothelial, and epithelial specific transcripts, a defining feature of mesothelial cells. Recent studies suggest that cells within the splanchnic mesothelium, an epithelial lining of the coelom, invade the adjacent splanchnic mesoderm and undergo an epithelial-to-mesenchymal transition at about the same time as the appearance of primitive endothelial and hematopoietic progenitors within the splanchnopleura [123]. These invading cells express cytokeratins, vimentin, and specific hemangioblastic markers. Primitive endothelial cells that develop in the splanchnopleura then colonize the floor of the aorta and differentiate in situ to produce the vasculature of the body wall, kidney, visceral organs, and limbs [124]. Several authors have argued that MSCs, which are delivered to the developing marrow cavity during endochondrial ossification by vascular invasion, may be derived directly from the endothelium [14, 125]. This hypothesis is supported by the fact that the antibody STRO-1 binds to the microvasculature of bone marrow [14] and is thought to react specifically with adventitial reticular cells, which function as specialized marrow pericytes. These pericytes have been shown to express alkaline phosphatase and α-smooth muscle actin and differentiate in situ into adipocytes to regulate the size and permeability of the marrow sinusoid system [14]. Post-capillary venule pericytes from bone marrow have also been shown to differentiate into cartilage and bone *in vivo* [126]. The similarity between MSCs and pericytes may indicate that some fraction of stromal elements arise from mesothelial-derived endothelial progenitors.

Our comparative studies also indicate that the MSC colony is uniquely characterized by its expression of the transcription factors *cbfa1* and *twist*. Expression of *cbfa1 in vivo* is required for bone formation and promotes osteoblast maturation [127]. Moreover, its expression is thought to specify a connective tissue cell fate to mesodermal cells during development [128]. Loss of the *twist* gene or its function in humans causes Saethre-Chotzen disease, characterized by craniosynostosis due to premature osteoblast differentiation [129, 130]. This is consistent with the fact that osteogenic cell differentiation has been shown to comprise a regulatory network involving FGF signalling, Twist, BMP2 and Id-1 [118]. In this scheme FGF signalling induces Twist expression, which blocks osteogenic commitment of mesodermal cells. BMP2 promotes osteogenic commitment by inducing the expression of Id-1, which in turn inhibits Twist. Since

Twist has also been shown to regulate *cbfa1* expression, its activity may play a pivotal role in the fate of MSCs.

5.2. The MSC population transcriptome

As indicated, the transcriptome of the MSC population contains fewer tags of unknown origin and a higher percentage of tags corresponding to skeletal and muscle specific transcripts as compared to the single cell-derived colony. These results imply that the population contains a higher percentage of committed progenitor/precursor cells, consistent with the fact that MSC populations are functionally heterogeneous and known to lose multi-potency with continued passage in culture [16, 83, 111]. Our comparative studies also showed that the population's transcriptome was significantly less unique than that of the colony when compared to other human cell types and tissues. This implies that the repertoire of expressed transcripts in the population more closely resembles that of the human cell types listed in Table 3.3. The MSC population was also characterized by many transcripts encoding proteins that mediate cell communication, neural activity, motility, and angiogenesis. Collectively, these findings may be interpreted to indicate that the transcriptome of the MSC population reflects the overall complexity of marrow stroma, the cells types that comprise it, and the functions they perform.

Ultra structural studies have revealed a complex architecture to the bone marrow stroma and a variety of morphologically distinct cell types within it [131]. For example, the major cell type forming the stroma of marrow is the reticular cell (RC), characterized by its sheet-like cytoplasmic processes that branch (reticulate) into the surrounding hematopoietic cords and interact directly with hematopoietic cells. The stroma also contains adventitial reticular (AR) and periarterial adventitial (PAA) cells. As indicated earlier AR cells line the walls of the venous sinuses forming an adventitial layer, and as such function as specialized marrow pericytes. PAA cells ensheath both arterioles and nerves, forming a functional barrier between the adventitia and hematopoietic parenchyma. In addition, bone marrow is also innervated by nervous tissue [132–134]. The majority of nerve fibers are distributed with blood vessels and are vasomotor. However, a significant fraction of myelinated and unmyelinated fibers are located between layers of PAA cells and within the hematopoietic cords [133].

It is well established that reticular (stromal) cells produce matrix proteins that form a fibrous meshwork throughout the bone marrow cavity, which maintains hematopoietic cells in an ordered arrangement during maturation and egress to the circulation. The latter process is mediated by AR cells via several mechanisms. First, AV cells may contract and/or physically migrate away from the sinus wall to permit transmural passage of hematopoietic cells. Motility of the AR cells is thought to be co-ordinated by communication with PAA and hematopoietic cells via the formation of gap junctions [131]. The formation of functional gap junctions between stromal cells and hematopoietic progenitor has also been demonstrated *in vitro* [135, 136]. Interestingly, osteoblasts have are thought to communicate via gap junctions to co-ordinate the remodelling of bone tissue [137]. The motility of AR cells also appears to be regulated by input from the nervous system. Efferent nerve terminals from fibers that track into the hematopoietic cords make attachments to basement membranes adjacent to PAA cells and/or terminate directly onto AR cells. Since the PAA and AR cells are connected by gap junctions, they are indirectly coupled to each other forming a circuit. This functional unit has been deemed the "neuro-reticular complex' and is thought to provide a means by which nervous input can alter stromal function to regulate hematopoiesis [133]. Lastly, AR cells also regulate the permeability of the sinuses by altering the amount of basement membrane they produce, which covers the basal endothelial surface [131].

Each of these processes, synthesis of basement membranes, cell communication via gap junctions, motility, formation of a physical barrier between cellular compartments, and responsiveness to neural activity are adequately represented by transcripts expressed in the MSC population. Therefore, the complexity (and apparent heterogeneity) of its transcriptome appears to reflect the summed activity of stromal elements in bone marrow.

As indicated previously, the MSC population also expressed neurotrophins, neurite inducing factors, and various angiogenins. The neurotrophin NGF has been shown to induce proliferation of hematopoietic precursors [138]. More recent findings indicate that AR cells express all three neurotrophin receptors [139]. Therefore, neurotrophins production by stromal cells may regulate hematopoiesis via an autocrine mechanism. Neurotrophins, axon guidance molecules, and neurite inducing factors produced by stromal cells also likely maintain nerve fibers in marrow and bone and guide their innervation during growth and remodelling after injury. Stromal cell production of angiogenic factors may serve a similar role in regulating angiogenesis, which is essential for bone formation and growth [140]. Recently, VEGF has been shown to induce expression of BMP4 in endothelial cells within the fracture microenvironment [141], as well as synergize with BMP4 to enhance bone healing [142]. The formation of anastomosing vascular structures in human long-term bone marrow cultures provides further evidence for the angiogenic activity of stromal elements [143].

In closing, it is fitting to comment on the growing number of reports suggesting that sub populations of nestin expressing cells in stroma represent neural stem/progenitor cells [101–104, 144, 145]. While nestin is expressed in the developing neuroepithelium of the central nervous system [146] it is also expressed in developing skeletal muscle, specifically in limb bud mesenchyme [147], in dermatomal cells and myoblasts during the earliest stages of myogenesis [149] and is up regulated in myoblasts during skeletal muscle regeneration [150]. Nestin expression has also been recently localized within endothelium of neonatal and adult vascular structures [151]. Therefore, expression of nestin may reflect the commitment of MSCs toward a myogenic fate or their derivation from endothelial cells. Interestingly, while various groups have evaluated the potential of nestin expressing MSCs to differentiate into neurons, no studies have evaluated their myogenic potential. Similarly, GFAP expression is not restricted to neural tissue, but also occurs in myxoid connective tissue and chondrocytes of elastic and fibrous cartilage as well as in myoepithelial cells and fibroblasts of ligamentum flavum and cardiac valves [152, 153]. Therefore, these intermediate filaments may perform specific but as of yet undefined functions in mesenchymal cells, including MSCs.

6. CONCLUDING REMARKS

Past studies profiling gene expression in MSCs have typically revealed a preponderance of expressed transcripts encoding structural proteins common in skeletal tissue and secreted factors that regulate hematopoiesis. Our SAGE analysis of the MSC transcriptome corroborated many of these previous findings, but also revealed that the cells express a plethora of transcripts encoding proteins involved in cell communication, motility, neural activity, angiogenesis and other biological processes that characterize marrow stroma. Therefore, the heterogeneity of the MSC transcriptome appears to reflect the nature and function of the different cell types that comprise this organ, the complexity of which is under appreciated. Our studies also indicate that the transcriptome of a clonal MSC population is characterized by a high percentage of expressed transcripts encoding proteins of an indeterminate nature. These findings reflect a basic lack of knowledge regarding the biology of MSCs, which attributes to the difficulty in ascribing a molecular phenotype to these stem cells. Comparative genomics studies indicate that transcripts uniquely expressed in MSCs include transcription factors and signalling molecules involved in limb bud morphogenesis, thereby providing clues to the regulatory mechanisms governing self-renewal and lineage commitment of MSCs. Deciphering these molecular pathways will further our understanding of the nature and biology of this unique stem cell population.

REFERENCES

[1] Phinney DG. Building a consensus regarding the nature and origin of mesenchymal stem cells. *J Cell Biochem*. 2002;38(suppl):7–12.

[2] Friedenstein AJ, Chailakhjan RK, Lalykina KS. The development of fibroblast colonies in monolayer cultures of guinea-pig bone marrow and spleen cells. *Cell Tissue Kinet*. 1970;3:393–403.

[3] Owen ME, Cave J, Joyner CJ. Clonal analysis *in vitro* of osteogenic differentiation of marrow stromal CFU-F. *J Cell Sci*. 1987;87:731–738.

[4] Maniatopoulos C, Sodek J, Melcher AH. Bone formation *in vitro* by stromal cells obtained from bone marrow of young adult rats. *Cell Tissue Res*. 1988;254:317–330.

[5] Leboy PS, Beresford JN, Devlin C, Owen ME. Dexamethasone induction of osteoblast mRNAs in rat marrow stromal cell cultures. *J Cell Physiol*. 1991;146:370–378.

[6] Bennett JH, Joyner CJ, Triffitt JT, Owen ME. Adipocytic cells cultured from marrow have osteogenic potential. *J Cell Sci*. 1991;99:131–139.

[7] Beresford JN, Bennett JH, Devlin C, Leboy PS, Owen ME. Evidence for an inverse relationship between the differentiation of adipocytic and osteogenic cells in rat marrow stromal cell cultures. *J Cell Sci*. 1992;102:341–351.

[8] Gimble JM, Robinson CE, Wu XY, et al. Peroxisome proliferators-activated receptor-gamma activation by thiazolidinediones induces adipogenesis in bone marrow stromal cells. *Mol Pharm*. 1996;50:1087–1094.

[9] Johnstone B, Hering TM, Caplan AI, Goldberg VM, Yoo JU. *In vitro* chondrogenesis of bone marrow-derived mesenchymal progenitor cells. *Expt Cell Res*. 1998;238:265–272.

[10] Mackay AM, Beck SC, Murphy JM, Barry FP, Chichester CO, Pittenger ME. Chondrogenic differentiation of cultured human mesenchymal stem cells from marrow. *Tissue Eng*. 1998;4:415–428.

[11] Dennis JE, Merriam A, Awadallah A, Yoo JU, Johnstone B, Caplan AI. A quadripotential mesenchymal progenitor cell isolated from the marrow of an adult mouse. *J Bone Miner Res*. 1999;14:700–709.

[12] Pittenger MF, MacKay AM, Beck SC, et al. Multilineage potential of adult human mesenchymal stem cells. *Science*. 1999;284:143–147.

[13] Kuznetsov SA, Krebsbach PH, Satomura K, et al. Single-colony derived strains of human marrow stromal fibroblasts form bone after transplantation *in vivo*. *J Bone Miner Res*. 1997;12:1335–1347.

[14] Bianco P, Riminucci M, Gronthos S, Robey PG. Bone marrow stromal stem cells: nature, biology, and potential applications. *Stem Cells*. 2001;19:180–192.

[15] Phinney DG, Kopen G, Righter W, Webster S, Tremain N, Prockop DJ. Donor variation in the growth properties and osteogenic potential of human marrow stromal cells. *J Cell Biochem*. 1999;75:424–436.

[16] Mauraglia A, Cancedda R, Quarto R. Clonal mesenchymal progenitors from human bone marrow differentiate *in vitro* according to a hierarchical model. *J Cell Sci*. 2000;113:1161–1166.

[17] Tremain N, Korkko J, Ibberson D, Kopen GC, DiGirolamo C, Phinney DG. MicroSAGE analysis of 2353 expressed genes in a single cell-derived colony of undifferentiated human mesenchymal stem cells reveals mRNAs of multiple cell lineages. *Stem Cells*. 2001;19:408–418.

[18] Dexter TM, Allen TD, Lajtha LG. Conditions controlling the proliferation of hematopoietic stem cells *in vitro*. *J Cell Physiol*. 1997;91:335–344.

[19] Testa NG, Dexter TM. Long-term production of erythroid precursor cells (BFU) in bone marrow cultures. *Differentiation*. 1977;9:193–195.

[20] Williams N, Jackson H, Sheridan APC, Murphy MJ, Elste A, Moore MAS. Regulation of megakaryopoiesis in long-term murine bone marrow cultures. *Blood*. 1978;51:245–255.

[21] Whitlock CA, Witte ON. Long-term cultures of B-lymphocytes and their precursors from murine bone marrow. *Proc Natl Acad Sci USA*. 1982;79:3608–3612.

[22] Gartner S, Kaplan H. Long-term culture of human bone marrow cells. *Proc Natl Acad Sci USA*. 1980;77:4756–4759.

[23] Dexter TM, Wright EG, Krisza F, Lajtha LG. Regulation of heamopoietic stem cell proliferation in long-term bone marrow cultures. *Biomedicine*. 1977;27:344–349.

[24] Mauch P, Greenberger JS, Botnick L, Hannon E, Hellman S. Evidence for structured variation in self-renewal capacity within long-term bone marrow cultures. *Proc Natl Acad Sci USA*. 1980;74:3879–3882.

[25] Coulombel L, Eaves AC, Eaves CJ. Enzymatic treatment of long-term human marrow cultures reveals the preferential location of primitive hemopoietic progenitors in the adherent layer. *Blood*. 1983;62:291–297.

[26] Funk PE, Kincade PW, Witte PL. Native associations of early hematopoietic stem cells and stromal cells isolated in bone marrow cell aggregates. *Blood*. 1994;83:361–369.

[27] Siczkowski M, Clarke D, Gordon MY. Binding of primitive hematopoietic progenitor cells to marrow stromal cells involves heparin sulfate. *Blood*. 1992;80:912–919.

[28] Hassan HT, Sadovinkova EY, Drize NJ, Zander AR, Neth R. Fibronectin increases both non-adherent cells and CFU-GM while collagen increases adherent cells in human normal long-term bone marrow cultures. *Haematologia, Budap*. 1997;28:77–84.

[29] Guerriero A, Worford L, Holland HK, Guo G-R, Sheehan K, Waller E. Thrombopoietin is synthesized by bone marrow stromal cells. *Blood*. 1997;90:3444–3455.

[30] Talts JF, Falk M, Ekblom M. Expansion of the nonadherent myeloid cell population by monoclonal antibodies against tenascin-C in murine long-term bone marrow cultures. *Expt Hematol*. 1998;26:552–561.

[31] Gupta P, Oegema TR, Brazil JJ, Dudek AZ, Slungaard A, Verfaillie CM. Structurally specific heparin sulfates support primitive human hematopoiesis by formation of a multimolecular stem cell niche. *Blood*. 1988;91:4641–4651.

[32] Gordon MY, Riley GP, Watt SM, Greaves MF. Compartmentalization of a hematopoietic growth factor (GM-CSF) by glycosaminoglycans in the bone marrow microenvironment. *Nature*. 1987;326:403–405.

[33] Wilkins BS, Jones DB. Immunohistochemical characterization of intact stromal layers in long-term cultures of human bone marrow. *Brit J Hematol*. 1995;90:757–766.

[34] Lobb RR, Hemler ME. The pathophysiologic role of $\alpha4$ integrins *in vivo*. *J Clin Invest*. 1994;94:1722–1728.

[35] Saeland S, Duvert V, Caux C, et al. Distribution of surface-membrane molecules on bone marrow and cord blood $CD34^+$ hematopoietic cells. *Expt Hematol*. 1992;20:24–33.

[36] Liesveld JL, Winslow JM, Frediani KE, Ryan DH, Abboud CN. Expression of integrins and examination of their adhesive function in normal and leukemic hematopoietic cells. *Blood*. 1993;81:112–121.

[37] Kerst JM, Sander JB, Slaper-Cortenbach ICM, et al. $\alpha4\beta1$ and $\alpha5\beta1$ are differentially expressed during myelopoiesis and mediate the adherence of $CD34^+$ cells to fibronectin in an activation-dependent pathway. *Blood*. 1993;81:344–351.

[38] Möhle R, Murea S, Kirsch M, Haas R. Differential expression of L-Selectin, VLA-4 and LFA-1 on $CD34^+$ progenitor cells fro bone marrow and peripheral blood during G-CSF-enhanced recovery. *Expt Hematol*. 1995;23:1535–1542.

[39] Simmons PJ, Masinovsky B, Longenecker BM, Berenson R, Torok-Storb B, Gallatin WM. Vascular cell adhesion molecule-1 expressed by bone marrow stromal cells mediates the binding of hematopoietic progenitor cells. *Blood.* 1992;80:388–395.

[40] Ryan DH. Adherence of normal and neoplastic human B cell precursors to the bone marrow microenvironment. *Blood Cells.* 1993;19:225–241.

[41] Teixido J, Hemler ME, Greenberger JS, Anklesaria P. Role of beta1 and beta2 integrins in the adhesion of human CD34[hi] stem cells to bone marrow stroma. *J Clin Invest.* 1992;90:358–367.

[42] Miyake K, Weissman IL, Greenberger JS, Kincade PW. Evidence for a role of the integrin VLA-4 in lympho-hematopoiesis. *J Expt Med.* 1991;173:599–607.

[43] Miyake K, Medina K, Ishihara M, Kimoto R, Auerbach R, Kincade W. A VCAM-1 like adhesion molecule on murine bone marrow stromal cells mediates binding of lymphocyte precursors in culture. *J Cell Biol.* 1991;114:557–565.

[44] Yanai N, Sekine C, Yagita H, Obinata M. Roles for integrin very late activation antigen-4 in stroma-dependent erythropoiesis. *Blood.* 1994;83:2844–2850.

[45] Papayannopoulou T, Nakamoto B. Peripheralization of hematopoietic progenitors in primates treated with anti-VLA$_4$ integrin. *Proc Natl Acad Sci USA.* 1993;90:9374–9378.

[46] Miyake K, Medina KL, Hayashi S-I, Ono S, Hamaoka T, Kincade PW. Monoclonal antibodies to Pgp-1/CD44 block lympho-hemopoiesis in long-term bone marrow cultures. *J Exp Med.* 1990;171:477.

[47] Lisovsky M, Braun SE, Ge Y, et al. Flt-3 ligand production by human bone marrow stromal cells. *Leukemia.* 1996;10:1012–1018.

[48] Weimar IS, Miranda N, Muller EJ, et al. Hepatocyte growth factor/scatter factor (HGF/SF) is produced by human bone marrow stromal cells and promotes proliferation, adhesion, and survival of human hematopoietic progenitor cells (CD34[+]). *Expt Hematol.* 1998;26:885–894.

[49] Li L, Milner LA, Deng Y, et al. The human homolog of rat Jagged1 expressed by marrow stroma inhibits differentiation of 32D cells through interaction with Notch1. *Immunity.* 1998;8:43–55.

[50] Tordjman R, Ortega N, Coulombel L, Plouet J, Romeo PH, Lemarchandel V. Neuropilin-1 is expressed on bone marrow stromal cells: a novel interaction with hematopoietic cells? *Blood.* 1999;94:2301–2309.

[51] Zannettino AC, Buhring HJ, Niutta S, Watt SM, Benton MA, Simmons PJ. The sialomucin CD164 (MGC-24v) is an adhesive glycoprotein expressed by human hematopoietic progenitors and bone marrow stromal cells that serves as a potent negative regulator of hematopoiesis.*Blood.* 1998;92:2613–2628.

[52] Gray Parkin K, Stephan RP, Apilado RG, et al. Expression of CD28 by bone marrow stromal cells and its involvement in B lymphopoiesis. *J Immunol.* 2002;169:2292–2302.

[53] Saad-Barda M, Rozenszajn LA, Ashush H, Shav-Tal Y, Ben Nun A, Zipori D. Adhesion molecules involved in the interactions between early T cells and mesenchymal bone marrow stromal cells. *Expt Hematol.* 1999;27:834–844.

[54] Siler U, Rousselle P, Muller CA, Klein G. Laminin gamma2 chain as a stromal cell marker of the human bone marrow microenvironment. *Brit J Hemotol.* 2002;119:212–220.

[55] Kittler ELW, MacGrath H, Temeles D, Crittenden R, Kister K, Quesenberry P. Biologic significance of constitutive and subliminal growth factor production by bone marrow stroma. *Blood.* 1992;79:3168–3178.

[56] Deryugina EI, Muller-Sieburg CE. Stromal cells in long-term cultures: keys to the elucidation of hematopoietic development. *Crit Rev Immun.* 1993;13:115–150.

[57] Yamashita T, Takahashi S, Ogata E. Expression of activin A/erythroid differentiation factor in murine stromal cells. *Blood.* 1992;79:304–307.

[58] Iwata M, Graf L, Awaya N, Torok-Storb B. Functional interleukin-7 receptors (IL-7Rs) are expressed by marrow stromal cells: binding of IL-7 increases levels of IL-6 mRNA and secreted protein. *Blood.* 2002;100:1318–1325.

[59] Nagao T. Significance of bone marrow stromal cells in hematopoiesis and hematological disorders. *J Exp Clin Med.* 1995;20:121–130.

[60] Liesveld JL, Abboud CN, Duerst RE, Ryan DH, Brennan JK, Lichtman MA. Characterization of human marrow stromal cells: role in progenitor cell binding and granulopoiesis. *Blood.* 1989;73:1794–1800.

[61] Charbord P, Lerat H, Newton I, et al. The cytoskeleton of stromal cells from human bone marrow cultures resembles that of cultured smooth muscle cells. *Exp Hematol.* 1990;18:276–282.

[62] Perkins S, Fleishman RA. Stromal cell progeny of murine bone marrow fibroblast colony-forming units are clonal endothelial-like cells that express collagen IV and laminin. *Blood.* 1990;75:620–625.

[63] Hasthorpe S, Bogdanovski M, Rogerson J. Characterization of endothelial cells in murine long-term marrow culture. Implication for hemopoietic regulation. *Exp Hematol.* 1992;20:476–481.

[64] Filshie RJ, Zannettino AC, Makrynikola V, et al. MUC18, a member of the immunoglobulin superfamily, is expressed on bone marrow fibroblasts and a subset of hematological malignancies. *Leukemia.* 1998;12:414–421.

[65] Lepri E, Delfino DV, Migliorati G, Moraca R, Ayroldi E, Riccardi C. Functional expression of Fas on mouse bone marrow stromal cells: upregulation by tumor necrosis factor-alpha and interferon gamma. *Expt Hematol.* 1998;26:1202–1208.

[66] Gronthos S, Simmons PJ, Graves SE, Robey PG. Integrin mediated interactions between human bone marrow stromal precursor cells and the extracellular matrix. *Bone.* 2001;28:174–181.

[67] Kortenjann M, Nehls M, Smith AJ, et al. Abnormal bone marrow stroma in mice deficient for nemo-like kinase, Nlk. *Eur J Immunol.* 2001;31:3580–3587.

[68] Kim GS, Hong JS, Kim SW, et al. Leptin induces apoptosis via ERK/cPLA2/cytochrome c pathway in human bone marrow stromal cells. *J Biol Chem.* 2003;278: 21920–21929.

[69] Grellier P, Yee D, Gonzalez M. Characterization of insulin-like growth factor binding proteins (IGFBP) and regulation of IGFBP-4 in bone marrow stromal cells. *Br J Hematol.* 1995;90:249–257.

[70] Satomura K, Derubeis AR, Fedarko NS, et al. Receptor tyrosine kinase expression in human bone marrow stromal cells. *J Cell Physiol.* 1998;177:426–438.

[71] Preston MR, el Haj AJ, Publicover SJ. Expression of voltage-operated Ca2+ channels in rat bone marrow stromal cells *in vitro. Bone.* 1996;19:101–106.

[72] Simmons PJ, Torok-Strob B. CD34 expression by stromal precursors in normal and human adult bone marrow. *Blood.* 1991;78:2848–2853.

[73] Shur I, Marom R, Lokiec F, Socher R, Benayahu D. Identification of cultured progenitor cells from human marrow stroma. *J Cell Biochem.* 2002;87:51–57.

[74] Baddoo M, Hill K, Wilkinson R, et al. Characterization of mesenchymal stem cells isolated from murine bone marrow by negative selection. *J Cell Biochem.* 2003;89:1235–1249.

[75] Simmons PJ, Torok-Storb B. Identification of stromal cell precursors in human bone marrow by a novel monoclonal antibody STRO-1. *Blood.* 1991;78:55–62.

[76] Gronthos S, Zannettino AC, Graves SE, Ohta S, Hay SJ, Simmons PJ. Differential cell surface expression of the STRO-1 and alkaline phosphatase antigens on discrete developmental stages in primary cultures of human bone cells. *J Bone & Min Res.* 1999;14:47–56.

[77] Dennis JE, Carbillet JP, Caplan AI, Charbord P. The STRO-1+ marrow cell population is multipotential. *Cells Tissues Organs.* 2002;170:73–82.

[78] Quirici N, Soligo D, Bossolasco P, Servida F, Lumini C, Deliliers GL. Isolation of bone marrow mesenchymal stem cells by anti-nerve growth factor receptor antibodies. *Expt Hematol.* 2002;30:783–791.

[79] Caneva L, Soligo D, Cattoretti G, De Harven E, Deliliers GL. Immuno-electron microscopy characterization of human bone marrow stromal cells with anti-NGFR antibodies. *Blood Cells, Molecules & Diseases.* 1995;21:73–85.

[80] Bruder SP, Ricalton NS, Boynton RE, et al. Mesenchymal stem cell surface antigen SB-10 corresponds to activated leukocyte cell adhesion molecule and is nvolved in osteogenic differentiation. *J Bone & Min Res.* 1998;13:655–663.

[81] Barry FP, Boynton RE, Haynesworth S, Murphy JM, Zaia J. The monoclonal antibody SH-2, raised against human mesenchymal stem cells, recognizes an epitope on endoglin (CD105). *Biochem Biophys Res Comm.* 1999;265:134–139.

[82] Barry F, Boynton R, Murphy M, Haynesworth S, Zaia J. The SH-3 and SH-4 antibodies recognize distinct epitopes on CD73 from human mesenchymal stem cells. *Biochem Biophys Res Comm.* 2001;289:519–524.

[83] Pittenger MF, Mbalaviele G, Black M, Mosca JD, Marshak DR. Mesenchymal stem cells. In: Koller MR, Palsson BO, Master JRW, eds. *Human Cell Culture.* Netherlands: Kluwer Academic Publishers; 2001:chap 9.

[84] Haynesworth SE, Baber MA, Caplan AI. Cytokine expression by human marrow derived mesenchymal rogenitor cells *in vitro*: effects of dexamethasone and IL-1α. *J Cell Physiol.* 1996;166:585–592.

[85] Cheng L, Qasba P, Vanguri P, Thiede MA. Human mesenchymal stem cells support megakaryocyte and pro-platelet formation from CD34(+) hematopoietic progenitors. *J Cell Physiol.* 2000;184:58–69.

[86] Zimmermann S, Voss M, Kaiser S, Kapp U, Waller CF, Mertens UM. Lack of telomerase activity in human mesenchymal stem cells. *Leukemia.* 2003;17:1146–1149.

[87] Banfi A, Bianchi G, Notaro R, Luzzatto L, Cancedda R, Quarto R. Replicative aging and gene expression in long term cultures of human bone marrow stromal cells. *Tissue Eng.* 2002;8:901–910.

[88] Bianchi G, Banfi A, Mastrogiacomo M, et al. Ex vivo enrichment of mesenchymal cell progenitors by fibroblast growth factor 2. *Exp Cell Res.* 2003;287:98–105.

[89] Shi S, Gronthos S, Chen S, et al. Bone formation by human postnatal bone marrow stromal stem cells is enhanced by telomerase expression. *Nat Biotech.* 2002;20:587–591.

[90] Simonsen JL, Rosada C, Serakinci N, et al.Telomerase expression extends the proliferative life-span and maintains the osteogenic potential of human bone marrow stromal cells. *Nat Biotech.* 2002;20:592–596.

[91] Jia L, Young MF, Powell J, et al. Gene expression profile of human bone marrow stromal cells: high-throughput expressed sequence tags sequencing analysis. *Genomics.* 2002;79:7–17.

[92] Minguell JJ, Erices A, Conget P. Mesenchymal stem cells. *Exp Biol Med.* 2001;226:507–520.

[93] Frank O, Heim M, Jakob M, et al.B Real-time quantitative RT-PCR analysis of human bone marrow stromal cells during osteogenic differentiation *in vitro. J Cell Biochem.* 2002;85:737–746.

[94] Dieudonne SC, Kerr JM, Xu T, et al. Differential display of human marrow stromal cells reveals unique mRNA expression patterns in response to dexamethasone. *J Cell Biochem.* 1999;76:231–243.

[95] Doi M, Nagano A, Nakamura Y. Genome-wide screening by cDNA microarray of genes associated with matrix mineralization by human mesenchymal stem cells *in vitro. Biochem Biophys Res Comm.* 2002;290:381–390.

[96] Locklin RM, Riggs BL, Hicok KC, Horton HF, Byrne MC, Khosla S. Assessment of gene regulation by bone morphogenetic protein 2 in human marrow stromal cells using gene array technology. *J Bone & Min Res.* 2001;16:2192–2204.

[97] Sekiya I, Vuoristo JT, Larson BJ, Prockop DJ. *In vitro* cartilage formation by human adult stem cellsfrom bone marrow stroma defines the sequence of cellular and molecular events during chondrogenesis. *Proc Natl Acad Sci USA.* 2002;99:4397–4402.

[98] Jaiswal RK, Jaiswal N, Bruder SP, Mbalaviele G, Marshak DR, Pittenger MF. Adult human mesenchymal stem cell differentiation to the osteogenic or adipogenic lineage is regulated by mitogen-activated protein kinase. *J Biol Chem.* 2000;275:9645–9652.

[99] Gregory CA, Singh H, Perry AS, Prockop DJ. The Wnt signalling inhibitor dickopf-1 is required for reentry into the cell cycle of human adult stem cells from bone marrow. *J Biol Chem.* 2003;278:28067–28078.

[100] Spees JL, Olson SD, Ylostalo J, et al. Differentiation, cell fusion, and nuclear fusion during *ex vivo* repair of epithelium by human adult stem cells from bone marrow stroma. *Proc Natl Acad Sci USA.* 2003;100:2397–2402.

[101] Sanchez-Ramos J, Song S, Cardozo-Pelaez F, et al. Adult bone marrow stromal cells differentiate into neural cells *in vitro. Expt Neurol.* 2000;164:247–256.

[102] Woodbury D, Schwarz EJ, Prockop DJ, Black IB. Adult rat and human bone marrow stromal cells differentiate into neurons.*J Neurosci Res.* 2000;61:364–370.

[103] Woodbury D, Reynolds K, Black IB. Adult bone marrow stromal stem cells express germline, ectodermal, endodermal, and mesodermal genes prior to neurogenesis. *J Neurosci Res.* 2002;96:908–917.

[104] Kim BJ, Seo JH, Bubien JK, Oh YS. Differentiation of adult bone marrow stem cells into neuroprogenitor cells *in vitro. Neuroreport.* 2002;13:1185–1188.

[105] Vogel W, Grunebach F, Messam CA, Kanz L, Brugger W, Buhring HJ. Heterogeneity among human bone marrow-derived mesenchymal stem cells and neural progenitor cells. *Haematologica.* 2003;88:126–133.

[106] Wislet-Gendebien S, Leprince P, Moonen G, Rogister B. Regulation of neural markers nestin and GFAP expression by cultivated bone marrow stromal cells. *J Cell Sci.* 2003;116:3295–3302.

[107] Caplan AI. The mesengenic process. *Clin Plast Surg.* 1994;21:429–435.

[108] Majumdar MK, Thiede MA, Mosca JD, Moorman M, Gerson SL. Phenotypic and functional comparison of cultures of marrow-derived mesenchymal stem cells (MSCs) and stromal cells. *J Cell Physiol.* 1998;176:57–66.

[109] Velculescu CE, Zhang L, Vogelstein B, Kinzler KW. Serial analysis of gene expression. *Science.* 1995;270:484–487.

[110] Velculescu VE, Zhang L, Zhou W, et al. Characterization of the yeast transcriptome. *Cell.* 1997;88:243–251.

[111] DiGirolamo CM, Stokes D, Colter D, Phinney DG, Class R, Prockop DJ. Propagation and senescence of human marrow stromal cells in culture: a simple colony-forming assay identifies samples with the greatest potential to propagate and differentiate. *Brit J Hematol.* 1999;107:275–281.

[112] Datson NA, van der Perk-de Jong J, van den Berg MP, de Kloet ER, Vreugdenhll E. MicroSAGE: a modified procedure for serial analysis of gene expression in limited amounts of tissue. *Nuc AcidRes.* 1999;27:1300–1307.

[113] Laurin N, Brown JP, Morissette J, Raymond V. Recurrent mutation of the gene encoding sequestosome 1 (SQSTM1/p62) in Paget disease of bone. *Am J Hum Genet.* 2002;70:1582–1588.

[114] Blaschke RJ, Monaghan AP, Schiller S, et al. SHOT, a SHOX-related homeobox gene, is implicated in craniofacial, brain, heart, and limb development. *Proc Natl Acad Sci USA.* 1998;95:2406–2411.

[115] Thisse B, Stoetzel C, Gorostiza-Thisse C, Perrin-Schmitt F. Sequence of the twist gene and nuclear localization of its protein in endomesodermal cells of early Drosophila embryos. *EMBO J.* 1988;7:2175–2183.

[116] Fuchtbauer EM. Expression of M-Twist during postimplantation development of the mouse. *Dev Dyn.* 1995;204:316–322.

[117] Lee MS, Lowe GN, Strong DD, Wergedal JE, Glackin CA. TWIST, a basic helix-loop-helix transcription factor, can regulate the human osteogenic lineage. *J CellBiochem.* 1999;75:566–577.

[118] Rice DP, Aberg T, Chan Y-S, et al. Integration of FGF and TWIST in clavarial bone and suture development. *Development.* 2000;127:1845–1855.

[119] Newey SE, Howman EV, Ponting CP, et al. Syncoilin, a novel member of the intermediate filament superfamily that interacts with alpha-dystrobrevin in skeletal muscle. *J Biol Chem.* 2000;276:6645–6655.

[120] Merino R, Rodriguez-Leon J, Macias D, Ganan Y, Exonomides AN, Hurle JM. The BMP antagonist Gremlin regulates outgrowth, chondrogenesis, and programmed cell death in the developing limb. *Development.* 1999;126:5515–5522.

[121] Capdevila J, Tsukui T, Rodriquez E, Zappavigna V, Izpisua Belmonte JC. Control of vertebrate limb outgrowth by the proximal factor Meis2 and distal antagonism of BMPs by Gremlin. *Mol Cell.* 1999;4:839–849.

[122] Enver T, Greaves M. Loops, lineage and leukaemia. *Cell.* 1998;94:9–12.

[123] Munoz-Chapuli R, Perez-Pomares JM, Macias D, Garcia-Garrido L, Carmona R, Gonzalez M. Differentiation of hemangioblasts from embryonic mesothelial cells? A model on the origin of the veterbrate cardiovascular system. *Differentiation.* 1999;64:133–141.

[124] Pardanaud L, Luton D, Prigent M, Bourcheix L-M, Catala M, Dieterien-Lievre F. Two distinct endothelial lineages in ontogeny, one of the related to hematopoiesis. *Development.* 1996;122:1363–1371.

[125] Bianco P, Cossu G. Uno, nessuno e centomila: searching for the identity of mesodermal progenitors. *Expt Cell Res.* 1999;251:257–263.

[126] Diaz-Flores L, Gutierrez R, Gonzalez P, Varela H. Inducible perivascular cells contribute to the neochondrogenesis in grafted perichondrium. *Anat Rec.* 1991;229:1–8.

[127] Komori T, Yagi H, Nomura S, et al. Targeted disruption of Cbfa1 results in a complete lack of bone formation owing to maturational arrest of osteoblasts. *Cell.* 1997;89:755–764.

[128] Satomura K, Krebsbach P, Bianco P, Robey PG. Osteogenic imprinting upstream of marrow stromal cell differentiation. *J Cell Biochem.* 2000;78:391–403.

[129] WangJabs E. A TWIST in the fate of human osteoblasts identifies signalling molecules involved in skull development. *J Clin Invest.* 2001;107:1075–1077.

[130] Yousfi M. Lasmoles F, Lomri A, Delannoy P, Marie PJ. Increased bone formation and decreased osteocalcin expression induced by reduced Twist dosage in Saethre-Chotzen syndrome. *J Clin Invest.* 2001;107:1153–1161.

[131] Weiss L. The hematopoietic microenvironment of bone marrow: an ultrastructural study of the stroma in rats. *Anat Rec.* 1976;186:161–184.

[132] Miller ML, McCuskey RS. Innervation of bone marrow in the rabbit. *Scand J Hematol.* 1973;10:17–23.

[133] Yamazaki K, Allen TD. Ultrastructural morphometric study of efferent nerve terminals on murine bone marrow stromal cells, and the recognition of a novel anatomical unit: the "neuro-reticular complex." *Am J Anat.* 1990;187:261–276.

[134] Tabarowski Z, Gibson-Berry K, Felten SY. Noradrenergic and peptidergic innervation of the mouse femur bone marrow. *Acta Histochemica.* 1996;98:453–457.

[135] Rosendaal M, Gregan A, Green CR. Direct cell-cell communication in the blood forming system. *Tissue & Cell.* 1991;23:457–470.

[136] Durig J, Rosenthal C, Halfmeyer K, et al. Intercellular communication between bone marrow stromal cells and CD34+ hematopoietic progenitor cells is mediated by connexion 43-type gap junctions. *Brit J Hematol.* 2000;111:416–425.

[137] Civitelli R. Cellular metabolism in the human skeleton. Cell-cell communication in bone. *Calcif Tissue Int.* 1995;56(suppl 1):S29–S31.

[138] Matsuda H, Coughlin MD, Bienestock J, Denburg JA. Nerve growth factor promotes human hematopoietic colony growth and differentiation. *Proc Natl Acad Sci USA.* 1988;85:6508–6512.

[139] Labouyrie E, Dubus P, Groppi A, et al. Expression of neurotrophins and their receptors in human bone marrow. *Am J Path.* 1999;154:405–415.

[140] Gerber HP, Vu TH, Ryan AM, Kowalski J, Werb Z, Ferrara N. VEGF couples hypertrophic cartilage remodelling, ossification and angiogenesis during endochondral bone formation. *Nat Med.* 1999;5:623–628.

[141] Bouletreau PJ, Warren SM, Spector JA, et al. Hypoxia and VEGF up-regulate BMP-2 mRNA and protein expression in microvascular endothelial cells: implication for fracture healing. *Plastic & Recon Surg.* 2002;109:2384–2397.

[142] Peng H, Wright V, Usas A, et al. Synergistic enhancement of bone formation and healing by stem cell-expressed VEGF and bone morphogenetic protein-4. *J Clin Invest.* 2002;110:751–759.

[143] Wilkins BS, Jones DB. Vascular networks within the stroma of human long-term bone marrow cultures. *J Path.* 1995;177:295–301.

[144] Jia Y, Yang Y, Zhou Y, et al. Differentiation of rat bone marrow stromal cells into neurons induced by baicalin. *Chinese Med J.* 2002;82:1337–1341.

[145] Safford KM, Hicok KC, Safford SD, et al. Neurogenic differentiation of murine and human adipose-derived stromal cells. *Biochem Biophys Res Comm.* 2002;294:371–379.

[146] Lendahl U, Zimmerman LB, McKay RD. CNS stem cells express a new class of intermediate filament protein. *Cell.* 1990;60:585–595.

[147] Wroblewski J, Engstrom M, Edwall-Arvidsson C, Sjoberg G, Sejersen T, Lendahl U. Distribution of nestin in the developing mouse limb bud *in vivo* and in micro-mass cultures of cells isolated from limb buds. *Differentiation.* 1997;61:151–159.

[148] Kachinsky AM, Dominov JA, Miller JB. Myogenesis and the intermediate filament protein, nestin. *Dev Biol.* 1994;165:216–228.

[149] Vaittinen S, Lukka R, Sahlgren C, et al. The expression of intermediate filament protein nestin as related to vimentin and desmin in regenerating skeletal muscle. *J Neuropath Expt Neurol.* 2001;60:588–597.

[150] Mokry J, Nemecek S. Angiogenesis of extra- and intraembryonic blood vessels is associated with expression of nestin in endothelial cells. *Folia Biol.* 1998;44:155–161.

[151] Egerbacher M, Krestan R, Bock P. Morphology, histochemistry, and differentiation of the cat's epiglottic cartilage: a supporting organ composed of elastic cartilage, fibrous cartilage, myxoid tissue, and fat tissue. *Anat Rec.* 1995;242:471–482.

[152] Hainfellner JA, Voigtlander T, Strobel T, et al. Fibroblasts can express glial fibrillary acidic protein (GFAP) *in vivo. J Neuropath Exp Med.* 2001;60:449–461.

CHAPTER 4

IN VIVO HOMING AND REGENERATION OF FRESHLY ISOLATED AND CULTURED MURINE MESENCHYMAL STEM CELLS

R. E. PLOEMACHER

*Department of Hematology, Erasmus Medical Centre, Rotterdam,
The Netherlands*

1. INTRODUCTION

Bone marrow (BM)-derived mesenchymal stem cells (MSCs) have the potential to differentiate along various mesenchymal lineages including those forming bone, cartilage, tendon, fat, muscle and marrow stroma that supports hematopoiesis. In addition, MSC are held to have extensive proliferative potential, which may contribute to regenerative processes *in vivo*. MSC are routinely quantitated using the CFU-F (colony-forming unit fibroblast) assay wherein stromal progenitors form myofibroblast colonies of at least 50 cells in about 10 days of culture. It is not known whether the CFU-F represents the elusive stromal stem cell, or is in fact a more restricted stromal progenitor analogous to the CFU-C in the hematopoietic hierarchy. Unfortunately, there exist no repopulation assays for MSC analogous to what is used to operationally define the various hematopoietic stem cell (HSC) subsets. Both their assumed regenerative and differentiation potential make MSCs candidates for cell-based therapeutic strategies for mesenchymal tissue injuries and for hematopoietic disorders by both local and systemic application.

It is evident that locally grafted MSC, either as a single cell suspension or attached to a porous (natural or artificial) substrate, have a fair chance to survive in the host and contribute to the anatomy and possibly also the physiology of the hosts' tissues. This may be even more likely if there is a (local) need for regeneration of cell types that the grafted MSC can produce. However, when cells are grafted systemically they will have to find both a location and a way to leave the bloodstream and migrate through the vessel wall and organ-specific structures to their preferred niches. This situation may represent difficulties that are hard to surmount if the cells to be grafted were isolated in a state that did not match that of a migratory cell. Specifically, they may not be able

J.A. Nolta (ed.), Genetic Engineering of Mesenchymal Stem Cells, 81–92.

to recognize the environmental guidance cues, or to sense and react adequately to such molecules with their cellular motility machinery, expression of chemokine receptors and adherence molecules in order to efficiently migrate and transmigrate across vessel walls. These cells then will be dislocated and subjected to clearing mechanisms such as phagocytosis.

In the following we will discuss the limited reports on the homing and regeneration of MSC following their systemic grafting into various species. In addition, we will present data on the repopulation ability of MSC as compared to HSC in consecutive recipients.

1.1. Transplantation of MSC

While recent studies describe the beneficial effects of stromal cell/MSC infusion in different models of deficiency [1], MSC homing has not been studied systematically. In most studies, few donor cells were detected after their transplantation and it is often unclear what the phenotype of the detected donor cells was. The majority of these studies lack sufficient quantitation of the percentage, kinetics and distribution over organs and tissues of MSC that home to particular body sites after their transplantation, and no comparison is made of the routes of transplantation (i.v., i.p., locally). In the majority of studies, "homing" has not been studied as a tool to determine the percentage of infused MSC that seeded into specific organs or sites, but rather to obtain information about the net effect of seeding, survival and regeneration, or even solely to detect any donor stromal cells contribution.

In the first experiments reported, infused MSC could not be traced *in vivo* due to a lack of usable markers, and thus stromal layers or CFU-F derived colonies were grown from recipient organs in order to obtain predominantly qualitative information of any donor MSC contribution. Using these techniques some investigators, including our own laboratory, could demonstrate donor MSC homing and persistence *in vivo* [2, 3] while others were unable to do so [for example, 4–6]. Using genemarked cells in a murine transplantation model, donor cells were detected by the presence of the transgene in a broad spectrum of tissues [7, 8]. Typically however, the injected donor MSC could not be detected by sensitive PCR at 1 week following transplantation, but at 1–5 months post-transplant progeny of the donor MSCs accounted for 1–12% of the cells in BM, spleen, bone, lung and cartilage [9]. Similarly, following their in utero transplantation in sheep, cultured human MSC engrafted and persisted in multiple tissues for as long as 13 months and underwent site-specific differentiation into chondrocytes, adipocytes, myocytes and cardiomyocytes, BM stromal cells and thymic stroma [10]. In contrast, EGFP-expressing cultured baboon stromal cells were exclusively found in the BM of part of the infused Baboons, suggesting that BM-derived MSC preferentially home to the BM [11]. Donor cells were both found in a radiation-conditioned and in an unconditioned baboon, which suggests that MSC homing occurs independent of hematopoietic cell depletion. Further, cells from a cloned mouse stromal cell line expressing B-gal could be traced back for several weeks in the bone of the infused mice and were show to retain their osteogenic properties [12]. Real-time imaging of infused [111]Indium-oxine labelled cultured rat MSC showed a predominance of radioactivity accumulation in liver and lungs, and only about 1% of injected activity in the total bone/BM mass [13] suggesting sequestration of these cells in the body's capillary network. In another study, i.v. injected cultured rat MSC were predominantly retrieved in the lung, but also in kidneys and liver [14].

Routine i.v. transplantation of human stromal cells into immuno-deficient NOD/SCID mice and their long-term expression of transgenes has been reported [15, 16], however, it is not clear from all of these studies whether the cloned stromal cells still had multipotential MSC characteristics, to what tissues the stromal cells had homed and in which numbers, and whether they actually expanded their numbers locally.

There are only very few reports on MSC grafting in the clinical transplant setting, and it has been difficult to study MSC homing as this was precluded by ethical considerations [1] or because no donor cells could be detected [17–20]. In most of these studies total BM was infused rather than a homozygous MSC population. In two studies where patients received sex-mismatched allografts from HLA-matched or—mismatched family donors, donor cell containing stromal layers were retrieved in 14 out of 41 patients [21] and 4 out of 13 patients [22]. It should be noted that the donor fibroblastoid component in these stromal layers could have been formed by non-MSC as total (T-cell depleted) BM cells were grafted.

All of these studies together suggest that in specific animal models (a) MSC can be transplanted and detected long-term *in vivo*, but often not in the first week following grafting, (b) that there is contrasting data on defective homing or engraftment of MSC and (c) that MSC exhibit site-specific differentiation. In men the homing properties and long-term reconstitutive capacity of primary and culture-expanded MSCs remain largely unknown.

1.2. Factors involved in MSC homing

We have recently studied the effects of irradiation conditioning and MSC culture on homing of MSC in a syngeneic mouse model [23] and taken advantage of the possibility to discriminate EGFP-marked donor CFU-F derived colonies from wildtype CFU-F colonies under the fluorescence microscope. Twenty-four hours after i.v. transplantation of uncultured EGFP-transgenic BM cells into 7 Gy (sublethally) irradiated wildtype mice, as many as 55–65% of infused CFU-F were recovered from the total BM and 3.5–7% from the spleen. These observations fully corroborated with our earlier studies using chromosome-tagged cells [3] and show that BM-derived primary CFU-F display a highly efficient homing to BM and spleen.

Some irradiation conditioning (3 or 7 Gy total body) before CFU-F transplantation increases their 24-hour homing in the BM and spleen with a log [23]. An explanation for this observation could be that CFU-F niches have to be "emptied," or that the BM and spleen environment directly after irradiation provide signals that support the entry and 24-hour survival of grafted CFU-F. However, these assumptions are less likely as the CFU-F population is not one with a rapid turnover. Irradiation or treatment with some cytostatic agents does not directly affect the structural integrity of the stroma, but may rather inflict latent damage in stromal cells that much later becomes overt that is through defective CFU-F colony formation *in vitro* or severely limited regenerative capacity following their (ectopic) transplantation *in vivo* [24–27].

CFU-F are somewhat less radiosensitive to gamma irradiation than are HSC. As tested *in vivo*, their D_0 for gamma irradiation is 1.3–1.4 Gy [23], whereas their *in vitro* sensitivity for gamma radiation has been reported to be around 1.6 Gy [26] and 2.15–2.45 Gy for X-radiation [28, 29] CFU-F have been reported to be relatively more sensitive than hematopoietic cells to neutron irradiation [29]. Although the majority of primary BM-derived CFU-F home to the BM within 24 hours, their 24-hour homing ability decreased with 90% following only 24 hours of culture in the presence of various growth factors including SCF, TPO, EGF, PDGF-BB, FGF-1, 2- or 9, Flt3-L, alone or in combination. Primary CFU-F cultured for 48 hours, or cultured stromal cell lines with MSC differentiation abilities, could not at all be detected in lymphohematopoietic organs following their systemic grafting. These observations indicate that *in vitro* culture, even brief periods thereof, fully abrogate the BM CFU-F's ability to home to the BM and spleen. The mechanisms that lead to this loss of homing ability remain obscure. We speculate that many cultured CFU-F following their injection would be sequestrated in organs with extensive capillary and sinusoidal beds as liver and lung, but we could not retrieve many donor (EGFP$^+$) CFU-F from these organs at 24 hours after grafting (unpublished data). Although this suggests that the cells were dislodged in vessel systems elsewhere in the body, the possibility remains that directly after injection the grafted CFU-F were sequestrated in liver and lung, and possibly even BM proper, but in the course of hours were cleansed and destroyed by the phagocytic system.

We have observed excellent survival of lacZ and EGFP-marked stromal cells that were grown on cellulose acetate membranes *in vitro* and subsequently implanted i.p. or s.c. in syngeneic recipients (unpublished data). This indicates that immunological rejection of cultured MSC/stromal cells in a syngeneic model is unlikely to explain our inability to detect systemically infused stromal cells and briefly cultured primary MSC. Apparently, upon culture MSC undergo rapid changes in their adhesion molecule expression [30–32]. Unfortunately, any stringent characterization of murine MSC before and after culture is precluded by the fact that they cannot be isolated with high purity as their *in vivo* phenotype is incompletely characterized.

1.3. Trafficking and chemotaxis of MSC

Some research groups have presented data to support the notion that MSC are present in blood, although other investigators have not been able to find circulating MSC. As discussed in our preliminary results, we have been able to detect CFU-F in the blood of steady state mice and to induce CFU-F mobilization [33, 34]. In the human, controversy exists on the presence of MSC in steady state blood [35–38]. Also, after G-CSF treatment circulating MSC have been detected in patients with for example, breast cancer [39, 40], but their presence or fibroblastic nature have been severely questioned by others [41, 42]. In human a circulating fibrocyte has been described with the phenotype CD34+ CD45+ CD13+ and that is capable of synthesizing collagen [43]. The observation that the progeny of infused primary MSC can be traced back in the BM also suggests that circulating MSC may naturally exist. It has to be clarified what the phenotype and properties of circulating MSC/CFU-F is as compared to the BM-derived ones. Directed migration of stromal cells *in vivo* has been reported following local injection of human stromal cells. For instance, migration of cultured human marrow stromal cells along known pathways for migration of neural stem cells to successive layers of the brain has been observed 5 days following their direct injection into rat brain [44]. In addition, *in vitro* expansion and localization in calvarial sites was seen following subcutaneous transplantation in scid/scid mice [45]. Migration of donor BM stromal cells or their precursors to the murine thymus has been observed following transplantation of BM cells plus bone grafts, but not BM cells alone [46].

Various chemoattractant and chemokinetic factors for hematopoietic progenitors and stem cells have been described, including stromal-derived factor-1 (SDF-1), a ligand for the G-protein-coupled CXCR4 receptor [47, 48]. In the light of some observations that MSC/CFU-F can be found in the systemic circulation, there is reason to assume that these cells are subject to chemotactic stimuli as well, which may guide them to reallocate in the proper BM environment. In view of the existence of many chemokines and 7-membrane spanning G-protein coupled receptors it may be assumed that there will be specific chemoattractants, likely in synergy with cytokines, for MSC. Some specificity of chemokines is emerging recently. For instance, Macrophage Inflammatory Protein-3beta (MIP-3B) and its receptor CCR-7 have been indicated to be involved in migration of T and B lymphocytes and a small subset of CD34+ cells [49–51]. Similar to MIP-3B, Secondary Lymphoid tissue Chemokine (SLC) binds to CCR-7, and both agonists are mainly attractive for macrophage precursors. In our laboratory [52, 53] both 2-arachidonoylglycerol (2-AG, the endogenous ligand for the Cb2-R), and somatostatin (the ligand for the murine SST2-R), have also been implicated in chemotaxis of splenic B lymphocytes and hematopoietic progenitors, respectively. Chemotaxis of cells can be modulated by cytokines, and CD34+ cells have been reported to show enhanced SDF-1 directed chemotaxis by either pre-treatment or co-incubation with IL-3 or SCF [47, 54, 55]. Vascular smooth muscle cells have been reported to respond with activation of the Erk1/Erk2 MAP kinase activation pathway and enhanced migration to IL-3, PDGF or VEGF [56]. It has to be investigated whether these cytokines similarly are involved in the regulation of migration of BM-derived stromal cells, or MSC, as they all represent a stage in myofibroblast development and may have similar phenotypic features [57, 30]. The progeny of MSC, for example, BM stromal cells, can also produce factors with chemoattractive and chemokinetic properties, for example, SDF-1. The elaboration of SDF-1 by stromal cells *in vitro* can be modulated by cytokines as Flt3-L and Tpo [58].

2. EXPANSION OF MSC

2.1. Expansion of MSC in vitro

Although some investigators have studied the effect of a variety of cytokines on the CFU-F colony formation, less is known about their effect on CFU-F expansion *in vitro*. Platelet-derived growth factor-BB (PDGF) and epidermal growth factor (EGF) have been demonstrated to have the greatest ability to support human CFU-F colony growth in a dose-dependent manner in serum-free cultures [59, 60]. Simultaneous addition of PDGF and EGF had no effect on the number of colonies initiated but resulted in dose-dependent increases in mean colony diameter that were significant when compared with the effect of either factor alone or with the size of colonies elicited in control cultures by 20% FCS. A combination of SCF and IL-3 has been reported to moderately expand human CFU-F in serum-free suspension cultures [61]. Colter and colleagues [62] reported on a 10^9-fold expansion of human CFU-F over a 6 week culture period in the absence of added cytokines but using selected batches of fetal calf serum.

For murine CFU-F expansion the strongest stimulus was produced by PDGF and IL-3, dependent on the serum content of the cultures [63]. We have not observed any stimulatory effect of PDGF and EGF, alone or in combination, on expansion of murine CFU-F in serum-containing cultures over a 7-day period, neither did FGF-1 or FGF-2 support a numerical increase in CFU-F in such cultures. However, the combined presence of FGF-2 (but not FGF-1) and EGF led to a 2.5 fold expansion of CFU-F numbers in 7-day serum-containing cultures as measured by replating the cultured cells in a CFU-F assay (unpublished data).

2.2. Expansion and regeneration of MSC in primary recipients

CFU-F are first detected in the BM just before birth, and expand their numbers exponentially until 8 weeks after birth when their numbers are maintained at a plateau range between 1200–2000 per femur throughout life [64]. Spleen CFU-F are most numerous before birth (2000–3000 CFU-F/spleen) and decrease rapidly within the first 8 postnatal weeks down to about 180 CFU-F/spleen in which range they are found for the lifetime of the mouse.

From recent observations using tracking HSC methodology it can be concluded that conditioning of the recipient is not required for, or at least does not preclude, HSC homing proper [65, 66]. In contrast, subsequent HSC regeneration is determined by the extent of previous depletion of the HSC compartment, whereas the contribution of donor or host HSC in the regenerative process is a direct function of the competition between (residual) host HSC and effectively homed donor HSC [67]. Likewise, we have shown that regeneration of systemically grafted EGFP$^+$ CFU-F is greatly facilitated by prior radiation conditioning of the host [23]. Thus, we observed an abortive 10-fold expansion of donor CFU-F in the BM during the first week after transplant of primary CFU-F in unconditioned recipients, while no donor CFU-F were retrieved after day 14 post-transplantation. In contrast, in 3 and 7 Gy pre-irradiated recipients donor CFU-F expanded over 100-fold in the host BM during the first month post-transplant where they then constituted 50% and 90%, respectively, of the entire BM CFU-F compartment. Donor and host together cells normalized the size of the entire CFU-F compartment during this period.

2.3. Regenerative ability of MSC as tested through serial transplantation

In order to test whether CFU-F, or their precursors, possess extensive regenerative capacity as has been reported for HSC, we subjected BM cells from Ly5.2, ß-actin/EGFP-transgenic mice to multiple transplantation rounds. We injected full BM cells in order to include any precursor cell population of CFU-F. Briefly, one third of a femur content was i.v. injected into 7 Gy total body irradiated Ly5.1 (non-EGFP) recipient mice and at 1 day, or 5 and 10 weeks post-transplantation groups of these mice were killed and the femoral donor and host-type CFU-F content determined. We choose to condition the hosts sublethally as to avoid deaths of recipients due to transplant failure.

As the number of injected donor CFU-F was determined at day 0, and the number of homed donor CFU-F at day 1, we were able to calculate the expansion of donor-type CFU-F in a femur. Ly5.2 (donor-type) and Ly5.1 (host-type) BM cells were sorted and their CAFC-day35 frequency determined, so that the percentage of donor-type long-term repopulating HSC could also be calculated at these time points. At 5 and 10 weeks post-transplant the harvested BM cells from the primary recipient were again grafted i.v. into a secondary host in a dose of one third of a femur content. This procedure was repeated when retransplanting BM cells from secondary to tertiary hosts. The CAFC-day35 and CFU-F donor chimerism percentages observed at week 5 post-transplant were comparable in the first hosts, however, in the second and third hosts very few or no donor CFU-F were retrieved from the femurs indicating that reversion to recipient CFU-F type had occurred (Table 4.1). At the same time points we observed donor CAFC-day35 to be still present in the second and third hosts but their percentages gradually declined as previously reported by various investigators [68–70]. We did not find any significant differences when comparing the 5-week intervals with the 10-week intervals between the transplantations. Both at 5 and 10 weeks post-transplant the total CFU-F content of the bone marrow was within control values.

These observations suggest that the regenerative capacity of primary CFU-F (or their precursors) is more limited than that of HSC, and that CFU-F can essentially not regenerate the depleted BM CFU-F compartment in secondary hosts. Obviously, more experiments shall have to be performed to exclude the possibility that the time periods between the consecutive transplantations were to short, limiting the outgrowth of the CFU-F compartment in successive hosts. Our observations are remarkable, as they seem to contrast with ample observations on the *in vitro* expansion of HSC (absent or limited) or "MSC" (extensive) populations.

Table 4.1. Donor-type chimerism of primitive HSC (CAFC-day35) and MSC (CFU-F) in the bone marrow of three consecutive recipient groups at 5 weeks post-transplantation of one third of a femur.

Recipient	% Ly5.2+ (donor) CAFC-day35		% EGFP+ (donor) CFU-F	
	Expt 1	Exp 2	Exp 1	Exp 2
1st	63.2	74.5	73.0	71.6
2nd	18.9	21.3	0.9	0.7
3rd	5.2	1.3	0	0

3. DISCUSSION

The absence of a transplantation assay for MSC and the lack of extensive information about the *in vivo* behaviour of transplanted MSC have stimulated our interest in the fate of transplanted CFU-F, or their precursor cells, and their regenerative abilities in primary and subsequent recipients. The available data from literature and our group indicate that the majority of primary MSC/CFU-F from the BM homes to the BM sites within 24 hour following their systemic transplantation. This highly specific and efficient homing ability is largely lost upon even short *in vitro* culture periods. Most cultured BM CFU-F are sequestered in other organs or sites probably due to altered membrane phenotype and function and/or diminished migratory capacity. This circumstance leads to very ineffective transplantation efficacy and seems to severely limit clinical application of *in vitro* genetically engineered MSC. It would be interesting to investigate whether MSC isolated from different organs display organ-specific homing as primary BM-derived CFU-F/MSC do, and whether all of these MSC have similar differentiation properties. Such data could assist the physician in targeting systemically infused MSC to an organ. Published data suggest that the successfully homed donor MSC may repopulate various organs (for example, bone, cartilage, lung, BM, spleen) and display site-specific differentiation [10] and tissue-specific expression of genes [9].

The temporal character of the repopulation of host organs with the progeny of transplanted MSC presents a concern and suggests that the infused MSC are either proliferatively silenced, or have limited self-renewal ability. This ability forms the basis of the extensive long-term repopulating ability that is a hallmark of primitive HSC. Stem cells have classically been characterized by four criteria [71], that is (1) that they can undergo multiple, sequential self-renewing cell divisions, (2) that single stem cell-derived daughter cells can differentiate into more than one cell type, (3) that stem cells functionally repopulate the tissue of origin when transplanted in a damaged recipient, and (4) that stem cells even contribute differentiated progeny *in vivo* even in the absence of tissue damage, that is if sufficient donor cells are transplanted to compete significantly with the residual host cells. The available data suggest that freshly isolated MSC meet at least the first 3 criteria, and suggest that MSC are therefore true stem cells. Yet, our serial transplantation data demonstrate that MSC show rapid exhaustion and have less extensive self-renewal capacity as have HSC.

Whether HSC can proliferate without limit, or whether their regenerative capacity declines with repeated division, has been debated for decades. Prevailing opinion favours an intrinsic "decline." However, serial transfer experiments wherein specifically input and output of long-lived stem cells (long-term reconstituting cells, LTRCs) were monitored have challenged the view that expansion of passaged stem cells is limited by exhaustion, and indicate that augmentation after transplant is limited by extrinsic mechanisms whose effects are reversible either by further transfer of the stem cells into irradiated hosts or by administration of exogenous cytokines [72]. It will have to be further studied whether transplantation of more MSC as well administration of specific cytokines during the regenerative period following transplantation will abrogate the MSC exhaustion in secondary and tertiary recipient mice.

From our studies it does not become evident whether the transplanted CFU-F or their progeny are functionally active *in vivo*, as we have merely studied the ability of the injected BM cells to generate CFU-F in (consecutive) recipients. Such knowledge would require *in situ* study of marked MSC and their progeny. We have observed that following its depletion by the prior irradiation and donor BM cell transplantation the total CFU-F compartment (consisting of donor and host CFU-F) in the BM regenerated up to a normal pre-irradiation seize. This strongly suggests that the transplanted MSC are responsive to homoeostatic regulation mechanisms and therefore behave normally concerning this function.

4. CONCLUSIONS AND FUTURE WORK

Thirty years after the first reports by Friedenstein [73] we see a revived interest in MSC as pioneering studies have fired the expectation that MSC and their progeny will have broad application in many clinical fields including healing of musculoskeletal defects, coronary distress, diseases of the central nervous system, as well as immunoregulation and graft facilitation. The appealing clinical applications and concomitant commercial interests have spurred many studies, a number of which fall short of providing stringent evidence of MSC engraftment, regeneration, long-term (functional) engraftment and site-specific differentiation. Also the terminology used in this field is highly confusing and concealing. We are in need of more basic studies that may help directing the biotechnology field. As an example, the issue of whether the CFU-F assay is a direct measurement of MSC frequency is unresolved at present. In the absence of an *in vivo* MSC repopulation assay, we do not know whether MSC and CFU-F denote the same cell type, or should be regarded as subsequent hierarchical stages in the development of myofibroblasts, and cells with chondrogenic, osteogenic and adipogenic differentiation properties. When focussing more closely, we do not even fully understand what cell types are composing a single CFU-F derived colony as myofibroblastoid [57] and endothelial cell characteristics [74] have been recorded in addition to mRNAs characteristic for epithelial, neuronal, chondrocytic and osteogenic cells [75]. It is clear that we need better understanding of the MSC compartment in order to develop strategies for diffuse of site-specific delivery for therapeutic applications.

REFERENCES

[1] Horwitz EM, Prockop DJ, Gordon PL, et al. Clinical responses to bone marrow transplantation in children with severe osteogenesis imperfecta. *Blood*. 2001;97:1227–1231.

[2] Keating A, Singer JW, Killen PD, et al. Donor origin of the *in vitro* haematopoietic microenvironment after marrow transplantation in man. *Nature*. 1982;298:400–405.

[3] Piersma AH, Ploemacher RE, Brockbank KGM. Transplantation of bone marrow fibroblastoid stromal cells in mice via the intravenous route. *Br J Haematol*. 1983a;54:285–290.

[4] Friedenstein AJ, Ivonov-Smolenski AA, Chialakhyan RK, et al. Origin of bone marrow stromal mechanocytes in radiochimeras and heterotopic transplants. *Exp Hematol*. 1978;6:440–444.

[5] Wilson FD, Greenberg BR, Konrad PN, Klein AK, Walling PR. Cytogenetic studies on bone marrow fibroblasts from a male-female hematopoietic chimera. Evidence that stromal elements in human transplantation recipients are of host type. *Transplantation*. 1978;25:87–88.

[6] Golde DW, Hocking WG, Quan SG, Sparkes RS, Gale RP. Origin of human bone marrow fibroblasts. *Br J Haematol*. 1980;44:183–187.

[7] Hou Z, Nguyen Q, Frenkel B, et al. Evidence of peripheral blood-derived, plastic-adherent CD34(-/low) hematopoietic stem cell clones with mesenchymal stem cell characteristics. *Stem Cells*. 2000;1:252–260.

[8] Keating A, Berkahn L, Filshie R. A Phase I study of the transplantation of genetically marked autologous bone marrow stromal cells. *Hum Gene Ther*. 1998;9:591–600.

[9] Pereira RF, Halford KW, O'Hara MD, et al. Cultured adherent cells from marrow can serve as long-lasting precursor cells for bone, cartilage, and lung in irradiated mice. *Proc Natl Acad Sci U S A*. 1995;92:4857–4861.

[10] Liechty KW, MacKenzie TC, Shaaban AF, et al. Human mesenchymal stem cells engraft and demonstrate site-specific differentiation in utero transplantation in sheep. *Nature Medicine*. 2000;6:1282–1286.

[11] Devine SM, Bartholomew AM, Mahmud N, et al. Mesenchymal stem cells are capable of homing to the bone marrow of non-human primates following systemic infusion. *Exp Hematol*. 2001;29:244–255.

[12] Dahir GA, Cui Q, Anderson P, et al. Pluripotential mesenchymal cells repopulate bone marrow and retain osteogenic properties. *Clin Orthop*. 2000;379:134–145.

[13] Gao J, Dennis JE, Muzic RF, Lundberg M, Caplan AI. The dynamic *in vivo* distribution of bone marrow-derived mesenchymal stem cells after infusion. *Cells Tissues Organs*. 2001;169:12–20.

[14] Barbash IM, Chouraqui P, Baron J, et al. Systemic delivery of bone marrow-derived mesenchymal stem cells to the infarcted myocardium: feasibility, cell migration, and body distribution. *Circulation*. 2003;108:863–868.

[15] Nolta JA, Hanley MB, Kohn DB. Sustained human hematopoiesis in immunodeficient mice by cotransplantation of marrow stroma expressing human interleukin-3: analysis of gene transduction of long-lived progenitors. *Blood*. 1994;83:3041–3051.

[16] Brouard N, Chapel A, Thierry D, Charbord P, Peault B. Transplantation of gene-modified human bone marrow stromal cells into mouse-human bone chimeras. *J Hematother Stem Cell Res*. 2000;9:175–181.

[17] Simmons PJ, Przepiorka D, Thomas ED, Torok-Storb B. Host origin of marrow stromal cells following allogeneic bone marrow transplantation. *Nature*. 1987;328:429–432.

[18] Laver J, Jhanwar SC, O'Reilly RJ, Castro-Malaspina H. Host origin of the human hematopoietic microenvironment following allogeneic bone marrow transplantation. *Blood*. 1987;70:1966–1968.

[19] Agematsu K, Nakahori Y. Recipient origin of bone marrow-derived fibroblastic stromal cells during all periods following bone marrow transplantation in humans. *Br J Haematol*. 1991;79:359–365.

[20] Koc ON, Peters C, Aubourg P, et al. Bone marrow-derived mesenchymal stem cells remain host-derived despite successful hematopoietic engraftment after allogeneic transplantation in patients with lysosomal and peroxisomal storage diseases. *Exp Hematol*. 1999;27:1675–1681.

[21] Cilloni D, Carlo-Stella C, Falzetti F, et al. Limited engraftment capacity of bone marrow-derived mesenchymal cells following T-cell-depleted hematopoietic stem cell transplantation. *Blood*. 2000;96:3637–3643.

[22] Tanaka J, Kasai M, Imamura M, et al. Evaluation of mixed chimaerism and origin of bone marrow derived fibroblastoid cells after allogeneic bone marrow transplantation. *Br J Haematol*. 1994;86:436–438.

[23] Rombouts WJC, Ploemacher RE. Primary murine MSC show highly efficient homing to the bone marrow but lose homing ability following culture. *Leukemia*. 2003;17:160–170.

[24] Ploemacher RE, Brockbank KGM, Brons NHC, de Ruiter H. Latent sustained injury of murine hemopoietic organ stroma induced by ionizing irradiation. *Haematologica*. 1983;68:454–468.

[25] Piersma AH, Ploemacher RE, Brockbank KG. Radiation damage to femoral hemopoietic stroma measured by implant regeneration and quantitation of fibroblastic progenitors. *Exp Hematol*. 1983b;11:884–890.

[26] Nikkels PG, de Jong JP, Ploemacher RE. Radiation sensitivity of hemopoietic stroma: long-term partial recovery of hemopoietic stromal damage in mice treated during growth. *Radiat Res*. 1987a;109:330–341.

[27] Nikkels PG, de Jong JP, Ploemacher RE. Long-term effects of cytostatic agents on the hemopoietic stroma: a comparison of four different assays. *Leuk Res*. 1987b;11:817–825.

[28] Werts ED, Gibson DP, Knapp SA, DeGowin RL. Stromal cell migration precedes hemopoietic repopulation of the bone marrow after irradiation. *Radiat Res*. 1980;81:20–30.

[29] Meijne EIM, Ploemacher RE, Vos O, Huiskamp R. The effects of graded doses of 1 MeV fission neutrons or X-rays on the murine hematopoietic stroma. *Radiat Res*. 1992;131:302–308.

[30] Galmiche MC, Koteliansky VE, Briere J, Herve P, Charbord P. Stromal cells from human long-term marrow cultures are mesenchymal cells that differentiate following a vascular smooth muscle differentiation pathway. *Blood*. 1993;82:66–76.

[31] Deschaseaux F, Charbord P. Human marrow stromal precursors are alpha 1 integrin subunit positive. *J Cellular Physiology*. 2000;184:319–325.

[32] Deschaseaux F, Gindraux F, Saadi R, Obert L, ChalmersD, Herve P. Direct selection of human bone marrow mesenchymal stem cells using an anti-CD49a antibody reveals their CD45med, low phenotype. *British Journal of Haematology.* 2003;122:506–517.

[33] Piersma AH, Ploemacher RE, Brockbank KGM, Nikkels PGJ, Ottenheim CPE. Migration of fibroblastoid stromal cells in murine blood. *Cell Tissue Kinet.* 1985;18:589–595.

[34] Brockbank KGM, Ploemacher RE, van Peer CMJ. Splenic accumulation of stromal progenitor cells in response to bacterial lipopolysaccharides. *Exp Hematol.* 1983a;11:358–363.

[35] Zvaifler NJ, Marinova-Mutafchieva L, Adams G, et al. Mesenchymal precursor cells in the blood of normal individuals. *Arthritis Res.* 2000;2:477–488.

[36] Barry FP, Boynton RE, Haynesworth S, Murphy JM, Zaia J. The monoclonal antibody SH-2, raised against human mesenchymal stem cells, recognizes an epitope on endoglin (CD105). *Biochem Biophys Res Commun.* 1999;265:134–139.

[37] Koc ON, Gerson SL, Cooper BW, et al. Rapid hematopoietic recovery after coinfusion of autologous-blood stem cells and culture-expanded marrow mesenchymal stem cells in advanced breast cancer patients receiving high-dose chemotherapy. *J Clin Oncol.* 2000;18:307–316.

[38] Lazarus HM, Haynesworth SE, Gerson SL, Caplan AI. Human bone marrow-derived mesenchymal (stromal) progenitor cells (MPCs) cannot be recovered from peripheral blood progenitor cell collections. *J Hematother.* 1997;6:447–455.

[39] Fernandez M, Simon V, Herrera G, Cao C, Favero Del, Minguell JJ. Detection of stromal cells in peripheral blood progenitor cell collections from breast cancer patients. *Bone Marrow Transplant.* 1997;20:265–271

[40] Hehschler R, Junghahn I, Fichtner I, Becker M, Goan SR. Donor fibroblasts from human blood engraft in immunodeficient mice. *Blood.* 1998;92(suppl 1): S587a.

[41] Purton LE, Mielcarek M, Torok-Storb B. Monocytes are the likely candidate "stromal" cell in G-CSF-mobilized peripheral blood. *Bone Marrow Transplant.* 1998;21:1075–1076.

[42] Ojeda-Uribe M, Brunot A, Lenat A, Legros M. Failure to detect spindle-shaped fibroblastoid cell progenitors in PBPC collections. *Acta Haematol.* 1993;90:139–143.

[43] Chesney J, Bacher M, Bender A, Bucala R. The peripheral blood fibrocyte is a potent antigen presenting cell capable of priming naive T cells in situ. *Proc Natl Acad Sci U S A.* 1997;94: 6307–6312.

[44] Azizi SA, Stokes D, Augelli BJ, DiGirolamo C, Prockop DJ. Engraftment and migration of human bone marrow stromal cells implanted in the brains of albino rats – similarities to astrocyte grafts. *Proc Natl Acad Sci USA.* 1998;95:3908–3913.

[45] Oreffo RO, Virdi AS, Triffitt JT. Retroviral marking of human bone marrow fibroblasts: *in vitro* expansion and localization in calvarial sites after subcutaneous transplantation *in vivo. J Cell Physiol.* 2001;186:201–209.

[46] Li Y, Hisha H, Inaba M, et al. Evidence for migration of donor bone marrow stromal cells into recipient thymus after bone marrow transplantation plus bone grafts: a role of stromal cells in positive selection. *Exp Hematol.* 2000;28:950–960.

[47] Aiuti A, Webb IJ, Bleul C, Springer T, Gutierrez-Ramos JC. The chemokine SDF-1 is a chemoattractant for human CD34+ hematopoietic progenitor cells and provides a new mechanism to explain the mobilization of CD34+ progenitors to peripheral blood. *J Exp Med.* 1997;185:111–120.

[48] Moehle R, Bautz F, Rafii S, Moore MA, Brugger W, Kanz L. The chemokine receptor CXCR-4 is expressed on CD34+ hematopoietic progenitors and leukemic cells and mediates transendothelial migration induced by stromal cell-derived factor-1. *Blood.* 1998;91:4523–4530.

[49] Rossi DL, Vicari AP, Franz-Bacon K, McClanahan TK, Zlotnik A. Identification through bioinformatics of two new macrophage proinflammatory human chemokines: MIP-3alpha and MIP-3beta. *J Immunol.* 1997;158:1033–1036.

[50] Greaves DR, Wang W, Dairaghi DJ, et al. CCR6, a CC chemokine receptor that interacts with macrophage inflammatory protein 3alpha and is highly expressed in human dendritic cells. *J Exp Med.* 1997;186:837–844.

[51] Yoshida R, Imai T, Hieshima K, et al. Molecular cloning of a novel human CC chemokine EBI1-ligand chemokine that is a specific functional ligand for EBI1, CCR7. *J Biol Chem.* 1997;272:13803–13809.

[52] Jorda MA, Verbakel SE, Valk PJ, et al. Hematopoietic cells expressing the peripheral cannabinoid receptor migrate in response to the endocannabinoid 2-arachidonoylglycerol. *Blood.* 2002;99:2786–2793.

[53] Oomen SP, van Hennik PB, Antonissen C, et al. Somatostatin is a selective chemoattractant for primitive (CD34+) hematopoietic progenitor cells. *Exp Hematol.* 2002;30:116–125.

[54] Kim CH, Broxmeyer HE. *In vitro* behavior of hematopoietic progenitor cells under the influence of chemoattractants: stromal cell-derived factor-1, steel factor, and the bone marrow environment. *Blood.* 1998;91:100–110.

[55] Dutt P, Wang JF, Groopman JE. Stromal cell-derived factor-1 alpha and stem cell factor/kit ligand share signaling pathways in hemopoietic progenitors: a potential mechanism for cooperative induction of chemotaxis. *J Immunol.* 1998;161:3652–3658.

[56] Brizzi MF, Formato L, Dentelli P, et al. Interleukin-3 stimulates migration and proliferation of vascular smooth muscle cells: a potential role in atherogenesis. *Circulation.* 2001;103:549–554.

[57] Charbord P, Oostendorp R, Pang W, et al. Comparative study of stromal cell lines derived from embryonic, fetal, and postnatal mouse blood-forming tissues. *Exp Hematol.* 2002;30:1202–1210.

[58] Kusadasi N, Oostendorp RAJ, Uyterlinden DG, Dzierzak EA, Ploemacher RE. Flt3-L and Tpo modulate the SDF-1-mediated chemotactic activities of marrow and embryonic stromal cells. Submitted for publication.

[59] Hirata J, Kaneko S, Nishimura J, Motomura S, Ibayashi H. Effect of platelet-derived growth factor and bone marrow-conditioned medium on the proliferation of human bone marrow-derived fibroblastoid colony-forming cells. *Acta Haematol.* 1985;74:189–194.

[60] Gronthos S, Simmons PJ. The growth factor requirements of STRO-1-positive human bone marrow stromal precursors under serum-deprived conditions *in vitro. Blood.* 1995;85:929–940.

[61] Baksh D, Davies JE, Zandstra PW. Adult human bone marrow-derived mesenchymal progenitor cells are capable of adhesion-independent survival and expansion. *Exp Hematol.* 2003;31:723–732.

[62] Colter DC, Class R, DiGirolamo CM, Prockop DJ. Rapid expansion of recycling stem cells in cultures of plastic-adherent cells from human bone marrow. *Proc Natl Acad Sci U S A.* 2000;97:3213–3218.

[63] Wang QR, Yan ZJ, Wolf NS. Dissecting the hematopoietic microenvironment. VI. The effects of several growth factors on the *in vitro* growth of murine bone marrow CFU-F. *Exp Hematol.* 1990;18:341–347.

[64] Brockbank KGM, Ploemacher RE, van Peer CMJ. An *in vitro* analysis of murine hemopoietic fibroblastoid progenitors and fibroblastoid cell function during aging. *Mech Ageing Develop.* 1983b;22:11–21.

[65] Zhong JF, Zhan Y, Anderson WF, Zhao Y. Murine hematopoietic stem cell distribution and proliferation in ablated and nonablated bone marrow transplantation. *Blood.* 2002;100:3521–3526.

[66] Yoshimoto M, Shinohara T, Heike T, Shiota M, Kanatsu-Shinohara M, Nakahata T. Direct visualization of transplanted hematopoietic cell reconstitution in intact mouse organs indicates the presence of a niche. *Exp Hematol.* 2003;31:733–740.

[67] Down JD, Boudewijn A, Dillingh JH, Fox BW, Ploemacher RE. Relationships between ablation of distinct haematopoietic subsets and the development of donor bone marrow engraftment following recipient pretreatment with different alkylating drugs. *Br J Cancer.* 1994;70:611–616.

[68] Siminovitch L, Till JE, McCulloch EA. Decline in colony-forming ability of marrow cells subjected to serial transplantation into irradiated mice. *J Cell Physiol.* 1964;64:23–31.

[69] Ogden DA, Micklem HS. The fate of serially transplanted bone marrow cell populations from young and old donors. *Transplantation.* 1976;22:287–293.

[70] Wolf NS, Priestley GV, Averill LE. Depletion of reserve in the hemopoietic system: III. Factors affecting the serial transplantation of bone marrow. *Exp Hematol.* 1983;11:762–771.

[71] Verfaillie CM, Pera MF, Lansdorp PM. Stem cells: hype and reality. (American Society Hematolology Educational Program). *Hematology.* 2002;369–391. Review.

[72] Iscove NN, Nawa K. Hematopoietic stem cells expand during serial transplantation *in vivo* without apparent exhaustion. *Curr Biol.* 1997;7:805–808.

[73] Friedenstein AJ, Deriglasova UF, Kulagina NN, et al. Precursors for fibroblasts in different populations of hematopoietic cells as detected by the *in vitro* colony assay method. *Exp Hematol.* 1974;2:83–92.

[74] Perkins S, Fleischman RA. Stromal cell progeny of murine bone marrow fibroblast colony-forming units are clonal endothelial-like cells that express collagen IV and laminin. *Blood.* 1990;75:620–625.

[75] Tremain N, Korkko J, Ibberson D, Kopen GC, DiGirolamo C, Phinney DG. MicroSAGE analysis of 2,353 expressed genes in a single cell-derived colony of undifferentiated human mesenchymal stem cells reveals mRNAs of multiple cell lineages. *Stem Cells.* 2001;19:408–418.

CHAPTER 5

NON-HUMAN PRIMATE MODELS OF MESENCHYMAL STEM CELL TRANSPLANTATION

S. M. DEVINE[1] AND R. HOFFMAN[2]

Department of Medicine, Division of Oncology, Siteman Cancer Center, Washington University School of Medicine, St Louis, MO[1]; Department of Medicine, Section of Hematology/Oncology, University of Illinois at Chicago, Chicago, IL[2]

1. INTRODUCTION

A variety of animal model systems including the mouse, rat, fetal sheep, dog, pig, and non-human primate have been used to study the effects of mesenchymal stem cell (MSC) transplantation. Due to its phylogenetic proximity to humans, relatedness of its MHC complex, and a number of other practical considerations the juvenile olive baboon (papio anubis) provides a theoretically ideal model system for the study of MSC transplantation. However, there are a number of well recognized and significant limitations to this model, such as the high cost of maintaining transplanted animals, the lack of reagents cross reactive with baboon cells, and a host of other potential complications that have limited its applicability (Table 5.1). Despite these reservations, we ultimately decided to pursue studies in the baboon due to its obvious potential clinical relevance. In this chapter, we review several avenues of study in which the baboon model has been highly informative regarding the effects of MSC transplantation. Based on these studies, we have a better sense of the potential strengths and limitations of MSC as instruments for the correction or modification of human diseases.

2. PHENOTYPIC AND FUNCTIONAL SIMILARITIES BETWEEN HUMAN AND BABOON MSC

In pivotal studies performed by the group at Osiris Therapeutics, Inc., Pittenger et al. characterized human MSC as adherent non-hematopoietic cells (lacking expression of CD14, CD34, and CD45) identified by the monoclonal antibodies SH-2 (CD105),

J.A. Nolta (ed.), Genetic Engineering of Mesenchymal Stem Cells, 93–110.

Table 5.1. Advantages and limitations to studies using non-human primates.

Advantages	Limitations
Highly conserved MHC-complex	High cost
Phenotypic similarity to humans	Low number available for study
Similar responses to HGF as humans	Difficult to handle/breed large animals
Outbred species, similar to humans	Lack of antibodies/reagents
Robust, can withstand many manipulations	Difficulty defining MHC-compatibility between donor/recipient pairs
Similar response to radiation injury as humans	Lack of defined organ injury models
Phenotypic/functional similarities to human MSC	

MHC: Major histocompatibility complex, HGF: Hematopoietic growth factors, MSC: Mesenchymal stem cells.

SH-3 (CD73) and SH-4 (CD73) [1]. Human MSC were isolated from bone marrow aspirates as cells which grow in a plastic adherent layer following culture in low glucose medium supplemented with fetal bovine serum selected from lots previously identified to enhance the growth of mesenchymal progenitors [1, 2]. Human MSC were passaged and expanded extensively in culture (by a factor of at least 10^5) yet appeared to maintain a stable phenotype for at least 18 doublings [3]. The multipotent nature of these cells was demonstrated by their ability to form adipocytes, chondrocytes, and osteocytes under specific *in vitro* and *in vivo* conditions. This multipotentiality appears to be retained even after several weeks in culture. This work further extended the pioneering studies of both Friedenstein and Caplan who initially proposed the existence of a multipotential human mesenchymal precursor [4–6].

Additional characterization of human MSC has revealed they express integrins, matrix receptors, and secrete cytokines essential for the support of hematopoiesis [2–7]. Later studies have shown that the amount and type of human MSC available for expansion in culture may vary from individual to individual. A recent study suggests there might be two distinct populations of mesenchymal progenitors isolated from the plastic adherent layer which possess markedly different proliferative capacities [8]. In addition, there appears to be significant heterogeneity between normal individuals in the ability of MSC to expand in culture and to maintain multipotentiality [9–11]. Moreover, there is interindividual variability for bone marrow aspirates to expand into MSC which appears dependent on the site from which the marrow was recovered [9]. These factors were all considered during the design of studies in the baboon model.

Following the definition of the process for the isolation, culture, and expansion of human MSC, we reasoned the baboon would serve as a relevant large animal model if a process for identifying and expanding baboon MSC could be developed. We performed bone marrow aspirates on anesthetized baboons and found that using conditions essentially identical to those employed for isolating and expanding human MSC, we

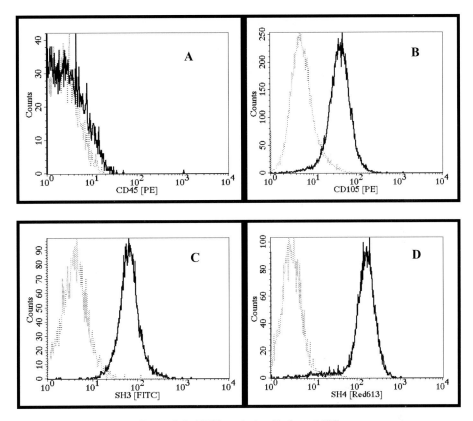

Figure 5.1. FACS analysis of baboon MSC.

could repeatedly collect, isolate, and expand plastic adherent baboon stromal cells. The stromal cells identified by the Osiris methods are both phenotypically and functionally similar to human MSC. Baboon MSC lack expression of CD34 and CD45, but stain positive for SH2 (CD105), SH3, and SH4 (CD73) (Figure 5.1). Functional similarities between human and baboon MSC were later demonstrated by the differentiation of baboon MSC into adipocytes and osteocytes using identical differentiation techniques as in humans [12]. We later demonstrated that when placed in an immunoisolatory device *in vivo* under high cell density conditions, baboon MSC were also able to differentiate into chondrocytes (Figure 5.2) [13].

Following the demonstration that baboon MSC were functionally and phenotypically similar to human MSC, the stage was set for a series of experiments addressing simple yet fundamental questions of MSC transplantation. These studies were mainly focused on answering questions of relatively immediate clinical relevance such as:

1. Are stromal cells capable of homing to the bone marrow following intravenous administration?

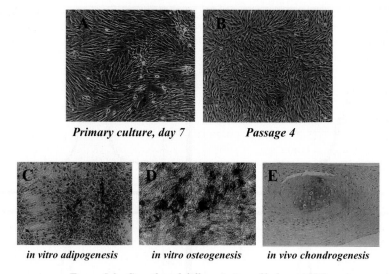

Primary culture, day 7   *Passage 4*

in vitro adipogenesis   *in vitro osteogenesis*   *in vivo chondrogenesis*

Figure 5.2. Growth and differentiation of baboon MSC.

2. How do MSC biodistribute following transplantation?
3. What are the effects of various routes of MSC administration?
4. What are the immunological consequences of transplanting MSC?
5. Are MSC potentially useful vehicles for gene therapeutic applications?

3. STUDIES OF MSC TRANSPLANTATION IN THE BABOON MODEL

3.1. Are stromal cells capable of homing to the bone marrow following intravenous administration?

In vitro studies performed nearly 30 years ago demonstrated that hematopoietic function was dependent upon contact with stromal cells [14–17]. The cellular components of the marrow stroma include macrophages, reticular endothelial cells, fibroblasts, adipocytes and osteogenic precursor cells which together provide growth factors, cell to cell interactions, and matrix proteins essential for the maintenance, growth, and differentiation of HSC [7, 14–22]. Recent studies have demonstrated that osteoblasts, progeny of MSC, play a critical supportive role within the bone marrow niche [22–24]. Functional defects in the marrow stroma would therefore be expected to impact negatively on hematopoiesis. For instance, SL^d mice, which bear a mutational defect in the membrane bound form of stem cell factor, the ligand for c-kit, render the marrow stroma nonfunctional and exhibit profound anemia at steady state. These mice are unable to support fully functioning hematopoiesis following stem cell transplantation [25–27]. Moreover, *in vitro* studies of both rodent and human bone marrow indicate that the marrow stromal elements are susceptible to damage by both radiation and chemotherapy [28–32]. Galloto et al. studied marrow stromal cell or progenitor cells on colony-forming units fibroblasts (CFU-F) in bone marrow transplant (BMT) re-cipients post transplant and compared these results to those of normal donors [33]. Marrow CFU-F frequencies were reduced by 60–90% (p < 0.05) and the numbers did not recover after more than 12 years following transplant. Stromal reconstitution to normal levels was found only in patients who were transplanted at ages less than 5 years old. Interestingly, patients with low CFU-F levels also had decreased bone density and significantly reduced levels of long term culture-initiating cells (LTC-IC). The cumulative evidence suggests that the recipient bone marrow stroma is frequently damaged following HSC transplantation and that such damage may potentially impair post transplant hematopoiesis.

The capacity to repair or replace stromal elements damaged by genetic mutation or exogenous influences may therefore be desirable. Theoretically, the transplantation of stromal cells from a healthy donor may improve either the overall rate of engraftment or pace of hematopoietic recovery following HSC transplantation. In SL^d mice, transplan-tation of stroma from normal mice corrects their anemia. Further, co-transplantation of donor whole bone grafts, osteoblasts, or stromal fibroblasts have been demonstrated to enhance engraftment of major histocompatibility complex (MHC)-mismatched bone marrow cells [34–37]. El-Badri et al. demonstrated that purified osteoblasts were capa-ble of enhancing engraftment of MHC mismatched HSC [37]. Almeida-Porada et al. demonstrated that co-transplantation of adult sheep stromal cells together with adult HSC resulted in increased donor cell engraftment in the bone marrow of fetal sheep and led to significantly increased levels of donor hematopoiesis for over 30 months following transplantation [36]. Kushida et al. demonstrated that the portal venous ad-ministration of murine whole bone marrow from normal allogeneic B6 donors into MRL/LPR mice resulted in full donor reconstitution and normalization of autoantibody levels [38]. Depletion of stromal cells from the whole bone marrow infusion abrogated these results, suggesting that marrow stromal cells mediate the graft enhancement. To-gether, the available evidence supports the concept that co-transplantation of stromal cells may enhance engraftment of HSC in a variety of settings. The mechanisms of enhancement, however, are not well understood.

Although stromal cell co-transplantation may be beneficial, the preponderance of evidence demonstrates that the bone marrow stroma remains of host origin following conventional HSC transplantation [39–45]. This has been confirmed by recent studies evaluating the origin of CFU-F or hMSC following allogeneic transplantation [46, 47]. To date, there has been no convincing clinical evidence that stromal elements derived from donors are capable of engraftment in the marrow (Table 5.2).

While the reasons for this are presently unclear, potential explanations include immune mediated rejection, inability to compete effectively with host stromal elements, or most likely the limited quantity of stromal cells contained within a typical allograft. Whether immune mediated rejection of donor stromal elements occurs is speculative and has not been demonstrated formally either in animals or the clinical setting. Studies by Horowitz et al. suggest that healthy stromal cells may be capable of engraftment if they are at a competitive advantage in comparison to an abnormal host stroma [48, 49]. In these studies, engraftment of donor-derived osteoblasts was detected following conventional bone marrow transplantation in two of three patients with osteogenesis

Table 5.2. Selected studies evaluating engraftment of stroma following hematopoietic stem cell transplantation.

Reference	Host origin	Donor origin	Comment
Keating [45]	Yes	Yes	Probable Contamination by donor derived macrophages
Simmons [39]	Yes	No	No donor chimerism when macrophages depleted from cultures
Laver [40]	Yes	No	Chromosomal analysis
Agematsu [43]	Yes	No	Fibroblast cultures
Santucci [41]	Yes	No	Fibroblast cultures
Horwitz [48]	Yes	Yes	1–2% donor derived osteoblasts detected in recipients with Osteogenesis imperfecta
Koc [47]	Yes	No	MSC cultures of post transplant marrow specimens
Galotto [33]	Yes	No	CFU-F cultures remained of host origin
Awaya [46]	Yes	No	Even as late as 27 years post-transplant, stroma remained of host origin
Cilloni [44]	Yes	Yes	Limited donor stromal chimerism detected in recipients of haploidentical stem cells

MSC: Mesenchymal stem cells, CFU-F: Colony forming units-fibroblast.

imperfecta. In most clinical situations, however, we favor the argument that the lack of detectable donor derived stroma is a function of the low numbers of MSC transplanted. It has been estimated that in humans, stromal progenitor cells are found in the bone marrow at a frequency of approximately 1 in 10^5 cells. Therefore, a typical bone marrow allograft may contain only 2–5×10^3 stromal progenitors/kg recipient weight. In addition, stromal progenitors are not detected to any significant degree in peripheral blood grafts [50]. Umbilical cord blood may contain MSC, but at very low frequency [51, 52]. If stromal cell engraftment is indeed dependent upon the number of stromal progenitors contained within a graft, increasing the number of MSC several fold should increase the likelihood of detection post-transplantation. We tested this hypothesis in the baboon.

We began by first culturing, isolating and expanding baboon MSC under conditions described for humans. During first passage, baboon MSC were retrovirally transduced with a vector that stably integrates into the genome, to provide a marker to track the engraftment of baboon MSC. In these experiments, five baboons were administered lethal radiation followed by intravenous autologous hematopoietic progenitor cells combined with either autologous (N = 3) or allogeneic (N = 2) mesenchymal stem cells following culture expansion. In four of these baboons, the mesenchymal stem cells were genetically modified by the retroviral vector encoding either the enhanced green fluorescent protein (N = 3), or the human placental alkaline phosphatase gene (N = 1) for tracking purposes. Another baboon received only intravenous GFP marked autologous MSC but no hematopoietic stem cells or conditioning irradiation. The experimental design using a total of six baboons is shown in Figure 5.3. The total MSC doses transplanted ranged from 3.6 to 30.0 × 10^6/kg although due to poor transduction efficiencies, lower numbers of gene marked cells were transplanted. Nevertheless, this represented a two- to threefold increase in the number of stromal cells typically transplanted with a conventional bone marrow graft, enabling us to address the primary hypothesis. First, we detected no acute or chronic toxicity associated with the intravenous infusion of expanded numbers of MSC. In all five recipients of gene marked MSC, transgene was detected in post-transplant bone marrow biopsies. In two animals receiving autologous MSC, including one non-conditioned recipient, transgene could be detected over 1 year following infusion. In one recipient of allogeneic gene marked MSC, transgene was detected in bone marrow at 76 days following infusion. These data were a definitive demonstration that when administered intravenously at sufficient quantities, MSC are capable of tracking back into the bone and bone marrow and have the capacity to re-establish residence there for an extended duration without causing toxicity [12]. Of note, while transgene was detected in bone marrow biopsies on a routine basis throughout the course of these experiments, bone marrow aspirates were typically negative for the presence of gene marked MSC. These observations may highlight a fundamental problem with the prior studies evaluating post-transplant stroma compartment chimerism. Following transplantation MSC or their progeny may

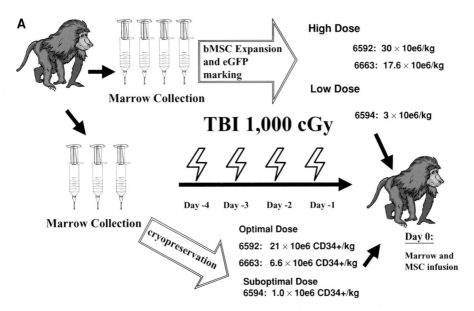

Figure 5.3.A. Autologous bone marrow and MSC transplant.

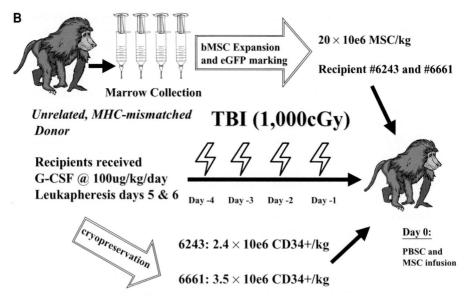

Figure 5.3.B. Autologous PBSC and MHC-mismatched MSC transplant.

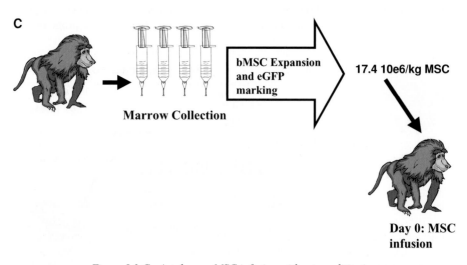

Figure 5.3.C. Autologous MSC infusion without conditioning.

lodge within the endosteal surface of the bone or migrate within compact bone rather than within the bone marrow cavity proper. Bone biopsies may be a better in situ method for analyzing the presence of transplanted MSC or their progeny whereas aspirates alone will not suffice for these purposes. Unfortunately, we have not yet developed good immunohistochemical methods in bone to detect actual engrafted cells and have had to rely on PCR-based assays. This is a serious limitation and allows us only to say that some of the transplanted cells made it back into the bone marrow and bones. We cannot rule out that the transplanted MSC fused with other cells resident within the marrow, as has been demonstrated previously with MSC and other adult somatic stem cells [53–58]. Nevertheless, although performed in a small number of animals, these findings do provide proof of principle that when given in large quantity, MSC are capable of engraftment within the bone and bone marrow microenvironments. We speculate that previous attempts to demonstrate donor stromal chimerism may have failed due to the low numbers of MSC contained within a typical allograft or to the techniques used for detection. We cannot provide any further information on the functional capacity of these cells once they have lodged within the bone. However, the limited numbers of cells that actually engrafted are unlikely to contribute substantially to the repair or regeneration of a damaged marrow stroma, at least not following intravenous administration.

3.2. What are the characteristics of MSC biodistribution following transplantation?

After intravenous administration it is unclear whether MSC home primarily to the bone and bone marrow microenvironments or if they distribute broadly (perhaps randomly) to a variety of non-hematopoietic tissues. Pereira and colleagues demonstrated that murine mesenchymal progenitors were capable of long term engraftment of bone, bone marrow, and lung following systemic administration [59]. Studies involving the tracking of genetically marked MSC have been undertaken in both dogs [60]. Short-term engraftment of allogeneic canine MSC expanded in culture and transduced with the enhanced green fluorescent protein (EGFP) gene was demonstrated following intravenous administration. Distribution studies suggested preferential homing to be bone and bone marrow, although lung, liver, and other tissues showed evidence of short-term MSC engraftment [60]. There was no evidence of toxicity or ectopic bone or cartilage deposition.

To address this question in the baboon, we performed long-term follow-up studies in three of the recipients that had been used in the studies described above. Two of these baboons had received lethal total body irradiation and hematopoietic support while one had not received any prior conditioning. Necropsies were performed between 9 and 21 months following MSC infusion and an average of 16 distinct tissues were recovered from each recipient and evaluated for the presence of GFP transgene within purified genomic DNA using a sensitive real-time polymerase chain reaction (PCR) assay. Two baboons had received autologous and one allogeneic MSC. Both allogeneic and autologous MSC distributed in a similar manner. Interestingly, gastrointestinal tissues harbored the highest concentrations of transgene/µg of DNA. Other tissues including kidney, lung, liver, thymus, and skin were also found to contain relatively high amounts of DNA equivalents. Estimated levels of engraftment in these tissues ranged from 0.1 to 2.7%. The non-conditioning recipient appeared to have less abundant engraftment. The results are shown in Figure 5.4 [61]. These findings suggest that MSC initially distribute broadly following systemic infusion and later may participate in ongoing cellular turnover and replacement in wide variety of tissues. The results also suggest

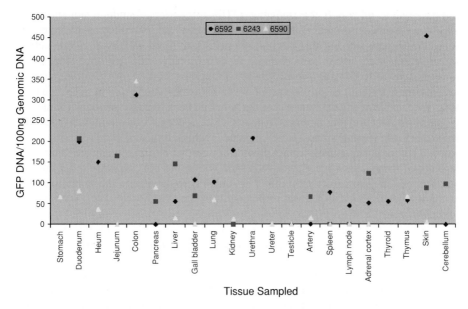

Figure 5.4. Long-term follow-up to assess engraftment levels in tissues from baboon recipients of MSC.
(A color version of this figure is freely accessible via the website of the book: *http://www. springer.com/1-4020-3935-2*)

that following systemic infusion the distribution of baboon MSC is not affected by histocompatibility or prior conditioning, although it is tempting to speculate that signals from the damaged gastrointestinal tissues attracted the intravenously administered MSC to distribute preferentially within the gastrointestinal tract. We cannot answer directly by these studies whether the MSC homed first into the bone marrow and then later redistributed or whether the tissues harbored MSC shortly following intravenous administration.

To further address the question of biodistribution of baboon MSC and to study the influence of conditioning and route of administration on the engraftment of MSC, we pursued a series of experiments in 11 additional baboons. Preliminary data suggested that MSC were capable of engraftment even if they were allogeneic to the recipient [12, 61]. Therefore, this set of studies used unrelated, MHC-mismatched baboon donors. MSC were isolated in culture and transduced using retroviral vectors encoding the neomycin resistant gene (NEOr) and either the human soluble tumor necrosis factor receptor (sTNFR) or the beta-galactosidase (bGAL) genes. Genetically modified MSCs were administered intravenously or directly injected into the bone marrow cavity (IBM) following 250cGy hemibody irradiation (HBI). In order to determine whether this dose of radiation injured the marrow stroma, we measured stromal progenitors (colony-forming unit-fibroblast CFU-F). There was no decrease in the CFU-F number following HBI with and without MSC grafts on either the irradiated or nonradiated sides of the recipients. By contrast, when four additional animals received an equivalent dose of *total* body irradiation (TBI) without MSC grafts, the numbers of CFU-F were reduced significantly. Following transplantation of baboon MSC by either route, bGAL positive cells were detectable for several months following transplantation in bone marrow aspirates obtained from animals receiving MSC grafts [62]. Levels of human soluble TNF receptor were also elevated in the serum of animals receiving MSC grafts. Neomycin resistant transgene was detected in the bone marrow of all animals who received MSC grafts by either intravenous or IBM route up to 6 months following transplantation. Only one of the eight animals had detectable copies of NEOr gene detected in multiple non-hematopoietic tissues as assessed by a quantitative real-time PCR assay 6 months following transplant. The level of radiation injury induced by 250cGy HBI did not appear to favor the localization of the implanted MSC to the irradiated side of the body. However, the studies confirmed that MSC grafts were capable of engrafting and persisting in the face of major immunological barriers. Genetically modified third party MSC were able to express and secrete an encoded protein for a prolonged period.

The reasons for the lower rates engraftment within the other non-hematopoietic tissues are difficult to explain, but may be secondary to the lower doses of MSC transplanted in comparison to the previous studies (30–50% of the doses previously transplanted), to the lack of lethal doses, or to the timing of necropsy (animals were necropsied at 6 months rather than at 9 to 21 months as in the first set of studies). Nevertheless, the ability of fully MHC-mismatched MSC to persist in this setting seems to suggest either that the transplanted MSC are not recognized by the host immune system or they possess the capacity to suppress local immune responses.

3.3. What are the immunological consequences of transplanting MSC?

The demonstration that genetically modified, MHC-disparate MSC resist prompt rejection by immunocompetent hosts suggests that MSC are not significantly immunogenic. Recent studies evaluating the immunomodulatory capacity of both human and baboon MSC suggest this may be the case. *In vitro* studies performed on human MSC were first reported by the group at Osiris Therapeutics [63]. They demonstrated that human MSC do not constitutively express Class II antigens or the T-cell co-stimulatory molecule B7. Further, human MSC are not substantially immunogenic and may actually inhibit both primary and secondary mixed lymphocyte reactions (MLR). Ongoing MLR reactions could also be suppressed by the addition of responder, stimulator, or intriguingly, third-party MSC. Using the baboon, we sought to determine whether these effects were operational *in vivo*. Bartholomew and colleagues first performed *in vitro* studies on baboon MSC to determine whether they were similar to human MSC [64]. Interestingly, baboon MSC failed to elicit a proliferative response from allogeneic lymphocytes. MSC added into a mixed lymphocyte reaction either on day 0 or on day 3 or to mitogen stimulated lymphocytes caused a greater than 50% reduction in the proliferative activity of the lymphocytes. This effect could be maximized by escalating the dose of MSC and could be reduced with the addition of exogenous interleukin-2. Next, the *in vivo* effects were tested in a stringent skin allograft rejection model. Baboon MSC were given by intravenous administration to MHC-mismatched recipient baboons prior to placement of autologous, donor, or third party skin grafts. Administration of MSC resulted in a modest but surprising prolongation of skin graft survival when compared to control

Table 5.3. Studies analyzing immunosuppressive effects of mesenchymal stem cells.

Reference	Species studied	Comments
Bartholomew [64]	Baboon	Donor, Host, and third party MSC equally inhibited T-cell proliferation in a dose dependent manner
DiNicola [88]	Human	Inhibitory effects of stromal cells on T-cell proliferation may be in part mediated through soluble factors such as HGF and TGF-β
Tse [71]	Human	Active inhibition of T-cell proliferation by MSC was not overcome by provision of co-stimulatory signals
Rasmussen [72]	Human	MSC were not targeted by cytotoxic T-cells of NK-cells. MSC did not inhibit NK cell mediated cytotoxicity
Potian [73]	Human	MSC exerted veto-like activity *in vitro*, but did not abrogate responses to recall antigens
Krampera [70]	Mouse	Data suggested murine MSC inhibit T-cell proliferation by physical hindrance of interaction between T-cells and antigen presenting cells
Kim [69]	Human/mouse	Human MSC injected into NOD/SCID mice increased engraftment of two MHC-disparate human umbilical cord blood grafts
Djouad [68]	Mouse	Murine MSC inhibited T-cell responses but also enhanced growth of melanoma cell line *in vivo*

animals ($11.3+0.3$ vs $7+/-0$ days, $p < 0.05$) [64]. These *in vivo* immunosuppressive effects are not trivial given that the modest skin allograft prolongation observed was similar to that obtained following conventional immunosuppressive agents such as cyclosporine or fludarabine [65, 66].

Since our report in the baboon system, the immunosuppressive effects of MSC have been confirmed by at least seven other groups using either murine or human cells (Table 5.3) [67–73]. DiNicola showed that human stromal cells could inhibit T-cell proliferation *in vivo* and that these effects may be mediated by transforming growth factor β1 or hepatocyte growth factor [67]. Krampera demonstrated that cell contact may be necessary for this effect and that it may be mediated thorough hindrance of T-cell interaction with antigen presenting cells [70]. Djouad used murine MSC to demonstrate that MSC could suppress allogeneic T-cell responses *in vivo* but that the effects were so profound that growth of injected B16 melanoma cells was promoted by co-injection of MSC [68]. Kim and colleagues used MSC co-transplantation into a NOD/SCID mouse UCB transplantation model to alleviate donor deviation and facilitate engraftment of multi-donor UCB [69].

Together, these findings should pave the way for studies to determine whether MSC can be used to reduce the amount of conditioning necessary for hematopoietic engraftment or to down modulate graft-versus-host responses. Further, as the immunogenicity of gene transduced hematopoietic stem cells has been raised in both pre-clinical and clinical gene therapy studies, MSC may also be useful vehicles in gene therapeutic applications [57, 74].

3.4. Are MSC potentially useful vehicles for gene therapeutic applications?

Given the limitations encountered in studies involving the genetic modification of HSC, recent attention has focused on the genetic manipulation of other somatic cells, including stromal cells and MSC. Since MSC proliferate extensively in culture, they are efficient targets for retroviral transduction [57, 75, 76]. The growth and differentiative potential of MSC does not appear to be affected by retroviral transduction [77, 78]. One strategy using genetically modified MSC introduces genes for secreted proteins into the MSC and then infuses them systemically so that they home into the bone marrow and secrete the therapeutic protein. In one study, autologous canine stromal cells transduced with the Factor IX gene were delivered systemically without toxicity [79]. Further, the systemic infusion of genetically modified MSC did not cause toxicity in the baboon [12]. A second strategy involves the encapsulation of protein secreting genetically modified MSC in some inert material that allows diffusion of protein but not the cells themselves. Human MSC transduced with the human erythropoietin (EPO) gene secrete bioactive erythropoietin and can correct drug induced anemia in NOD/SCID mice when placed in a subcutaneously implanted ceramic cube [80]. We next used the baboon system to evaluate MSC as cellular targets for gene therapy. In this set of experiments, baboon MSC were genetically modified with a bicistronic vector encoding the human EPO gene and the eGFP gene [81]. Transduction efficiencies ranged from 72 to 99% after incubation of MSC with retroviral supernatant. The transduced baboon MSC were capable of producing human erythropoietin *in vitro* before implantation. In order to determine the capacity of MSC to express human EPO *in vivo*, transduced MSC were injected intramuscularly into NOD/SCID mice. In a separate experiment, transduced MSC were loaded into immunoisolatory devices and surgically implanted into either autologous or allogeneic baboon recipients. Interestingly, human erythropoietin could be detected in the serum of NOD/SCID mice for up to 28 days and in the serum of five baboons from between 9 to 137 days. The NOD/SCID mice experienced sharp rises in hematocrit after intramuscular injection of the hEPO transduced MSC. The baboon that had expressed human erythropoietin for 137 days experienced a significant rise in hematocrit. The reason for the mechanisms underlying the loss of hEPO expression over time is unclear. Both gene silencing and promotor inactivation may play a role as the explanted cells were capable of producing human erythropoietin *in vitro* after explantation, but at markedly reduced levels compared to pre-implantation. Immunologic reactivity to the implanted MSC is unlikely since only one of the five baboon recipients developed any anti-donor antibodies and the immunoisolatory devices prevented a cellular immune response from infiltrating into the devices.

In another set of experiments, we transplanted MHC mismatched baboon MSC allografts transduced with the soluble TNF receptor at a dose of 5×10^6 per kilogram and measured levels of the soluble TNF receptor in the serum of the animals following transplantation [62]. Animals received hemibody irradiation and genetically modified MSC by either the IV or the IBM route. In four out of five evaluated animals, high levels of soluble TNF receptor were measured in the serum until day 10 posttransplant. Following this, levels declined progressively but remained elevated above control for 60 days in two out of five animals. Interestingly, the highest levels of soluble TNF receptor were noted in the serum of baboons that had received grafts by the IBM route. Together, these experiments suggest that genetically modified MHC-mismatched MSC are capable of expressing transgene and secreting a gene product without being rejected. Such findings are intriguing and suggest that gene-modified MSC may not be detected to any significant degree by the host immune system. Clearly, confirmatory studies will be necessary. However, the finding that genetically-modified MSC may engraft long term without requiring conditioning holds promise that MSC may be useful vehicles for somatic gene therapy without requiring the pre-conditioning which appears necessary for the durable engraftment of genetically modified HSC [82, 83]. Whether these findings can be further exploited in a clinically useful manner requires further study. Together, our preliminary studies using the baboon model suggest that genetically modified MSC may be useful for gene therapy purposes where the target cells are needed to secrete a gene product such as in lysosomal storage disease [84, 85]. MSC engineered to overexpress the pro-survival gene Akt were used to repair myocardial damage following experimentally induced infarction [86, 87]. Many studies of MSC-based gene therapy are anticipated in the ensuing years.

4. CONCLUSION

Mesenchymal stem cells hold promise as a form of cellular therapy for the repair or regeneration of damaged tissues. Moreover, the immuno-modulatory effects may also be exploited for recipients of HSC or solid organ transplantation. The pace of clinical trials has been slow to date since a number of fundamental questions regarding MSC biology remain unanswered. For instance, should MSC be delivered systemically, implanted, or infused into a particular target organ? What are the signals governing the growth and differentiation of MSC? What signals are required in order to attract MSC to a particular organ? Do MSC retain their multipotentiality following engraftment or do they undergo differentiation? If so, how will this affect their capacity to deliver therapeutic genes? What is the appropriate dose of MSC for transplantation and what are the best schedule and routes of administration? Are all MSC the same or is there heterogeneity among the population of cells we term MSC? To address these and many other questions, preclinical animal models will be critical. Obviously, the mouse is the most logical system to explore many of these questions given the capacity to manipulate them genetically and to generate reproducible results. Ultimately, it will be important to translate many of the encouraging findings from murine systems into the nonhuman primate model before the clinical potential of MSC therapy can be fully realized. Although resource intensive, expensive, and time consuming, the baboon has provided valuable insights into our understanding of MSC biology and the effects of the MSC transplantation.

REFERENCES

[1] Pittenger MF, Mackay AM, Beck SC, et al. Multilineage potential of adult human mesenchymal stem cells. *Science.* 1999;284(5411):143–147.

[2] Majumdar MK, Thiede MA, Mosca JD, Moorman M, Gerson SL. Phenotypic and functional comparison of cultures of marrow-derived mesenchymal stem cells (MSCs) and stromal cells. *J Cell Physiol.* 1998;176(1):57–66.

[3] Gerson SL. Mesenchymal stem cells: no longer second class marrow citizens [news; comment]. *Nat Med.* 1999;5(3):262–264.

[4] Friedenstein AJ, Deriglasova UF, Kulagina NN, et al. Precursors for fibroblasts in different populations of hematopoietic cells as detected by the *in vitro* colony assay method. *Exp Hematol.* 1974;2(2):83–92.

[5] Friedenstein AJ, Chailakhyan RK, Latsinik NV, Panasyuk AF, Keiliss-Borok IV. Stromal cells responsible for transferring the microenvironment of the hemopoietic tissues. Cloning *in vitro* and retransplantation *in vivo. Transplantation.* 1974;17(4):331–340.

[6] Caplan AI. Mesenchymal stem cells. *J Orthop Res.* 1991;9(5):641–650.

[7] Deans RJ, Moseley AB. Mesenchymal stem cells: biology and potential clinical uses. *Exp Hematol.* 2000;28(8):875–884.

[8] Colter DC, Class R, DiGirolamo CM, Prockop DJ. Rapid expansion of recycling stem cells in cultures of plastic-adherent cells from human bone marrow. *Proc Natl Acad Sci USA.* 2000;97(7):3213–3218.

[9] Digirolamo CM, Stokes D, Colter D, Phinney DG, Class R, Prockop DJ. Propagation and senescence of human marrow stromal cells in culture: a simple colony-forming assay identifies samples with the greatest potential to propagate and differentiate. *Br J Haematol.* 1999;107(2):275–281.

[10] Phinney DG, Kopen G, Isaacson RL, Prockop DJ. Plastic adherent stromal cells from the bone marrow of commonly used strains of inbred mice: variations in yield, growth, and differentiation. *J Cell Biochem.* 1999;72(4):570–585.

[11] Phinney DG, Kopen G, Righter W, Webster S, Tremain N, Prockop DJ. Donor variation in the growth properties and osteogenic potential of human marrow stromal cells. *J Cell Biochem.* 1999;75(3):424–436.

[12] Devine SM, Bartholomew AM, Mahmud N, et al. Mesenchymal stem cells are capable of homing to the bone marrow of non-human primates following systemic infusion. *Exp Hematol.* 2001;29(2):244–255.

[13] Bartholomew A, Patil S, Mackay A, et al. Baboon mesenchymal stem cells can be genetically modified to secrete human erythropoietin *in vivo. Hum Gene Ther.* 2001;12(12):1527–1541.

[14] Dexter TM. Stromal cell associated haemopoiesis. *J Cell Physiol Suppl.* 1982;1:87–94.

[15] Owen m. Lineage of osteogenic cells and their relationship to the stromal system. In: Peck WA, ed.. *Bone and Mineral Research.* London: Elsevier; 1985:1–25.

[16] Owen M. Marrow stromal stem cells. *J Cell Sci Suppl.* 1988;10:63–76.

[17] Owen M, Friedenstein AJ. Stromal stem cells: marrow-derived osteogenic precursors. *Ciba Found Symp.* 1988;136:42–60.

[18] Teixido J, Hemler ME, Greenberger JS, Anklesaria P. Role of beta 1 and beta 2 integrins in the adhesion of human CD34hi stem cells to bone marrow stroma. *J Clin Invest.* 1992;90(2):358–367.

[19] Deryugina EI, Muller-Sieburg CE. Stromal cells in long-term cultures: keys to the elucidation of hematopoietic development? *Crit Rev Immunol.* 1993;13(2):115–150.

[20] Bianco P, Gehron Robey P. Marrow stromal stem cells. *J Clin Invest.* 2000;105(12):1663–1668.

[21] Bianco P, Riminucci M, Gronthos S, Robey PG. Bone marrow stromal stem cells: nature, biology, and potential applications. *Stem Cells.* 2001;19(3):180–192.

[22] Lemischka IR, Moore KA. Stem cells: interactive niches. *Nature.* 2003;425(6960):778–779.

[23] Zhang J, Niu C, Ye L, et al. Identification of the haematopoietic stem cell niche and control of the niche size. *Nature.* 2003;425(6960):836–841.

[24] Calvi LM, Adams GB, Weibrecht KW, et al. Osteoblastic cells regulate the haematopoietic stem cell niche. *Nature.* 2003;425(6960):841–846.

[25] Anklesaria P, FitzGerald TJ, Kase K, Ohara A, Greenberger JS. Improved hematopoiesis in anemic Sl/Sld mice by splenectomy and therapeutic transplantation of a hematopoietic microenvironment. *Blood.* 1989;74(3):1144–1151.

[26] Anklesaria P, Kase K, Glowacki J, et al. Engraftment of a clonal bone marrow stromal cell line *in vivo* stimulates hematopoietic recovery from total body irradiation. *Proc Natl Acad Sci USA.* 1987;84(21):7681–7685.

[27] Kapur R, Majumdar M, Xiao X, McAndrews-Hill M, Schindler K, Williams DA. Signaling through the interaction of membrane-restricted stem cell factor and c-kit receptor tyrosine kinase: genetic evidence for a differential role in erythropoiesis. *Blood.* 1998;91(3):879–889.

[28] Chamberlin W, Barone J, Kedo A, Fried W. Lack of recovery of murine hematopoietic stromal cells after irradiation-induced damage. *Blood.* 1974;44(3):385–392.

[29] Fried W, Chamberlin W, Kedo A, Barone J. Effects of radiation on hematopoietic stroma. *Exp Hematol.* 1976;4(5):310–314.

[30] Carlo-Stella C, Tabilio A, Regazzi E, et al. Effect of chemotherapy for acute myelogenous leukemia on hematopoietic and fibroblast marrow progenitors. *Bone Marrow Transplant.* 1997;20(6):465–471.

[31] Domenech J, Roingeard F, Binet C. The mechanisms involved in the impairment of hematopoiesis after autologous bone marrow transplantation. *Leuk Lymphoma.* 1997;24(3–4):239–256.

[32] O'Flaherty E, Sparrow R, Szer J. Bone marrow stromal function from patients after bone marrow transplantation. *Bone Marrow Transplant.* 1995;15(2):207–212.

[33] Galotto M, Berisso G, Delfino L, et al. Stromal damage as consequence of high-dose chemo/radiotherapy in bone marrow transplant recipients. *Exp Hematol.* 1999;27(9):1460–1466.

[34] Hashimoto F, Sugiura K, Inoue K, Ikehara S. Major histocompatibility complex restriction between hematopoietic stem cells and stromal cells *in vivo. Blood.* 1997;89(1):49–54.

[35] Almeida-Porada G, Flake AW, Glimp HA, Zanjani ED. Cotransplantation of stroma results in enhancement of engraftment and early expression of donor hematopoietic stem cells in utero. *Exp Hematol.* 1999;27(10):1569–1575.

[36] Almeida-Porada G, Porada CD, Tran N, Zanjani ED. Cotransplantation of human stromal cell progenitors into preimmune fetal sheep results in early appearance of human donor cells in circulation and boosts cell levels in bone marrow at later time points after transplantation. *Blood.* 2000;95(11):3620–3627.

[37] El-Badri NS, Wang BY, Cherry Good RA. Osteoblasts promote engraftment of allogeneic hematopoietic stem cells. *Exp Hematol.* 1998;26(2):110–116.

[38] Kushida T, Inaba M, Takeuchi K, Sugiura K, Ogawa R, Ikehara S. Treatment of intractable autoimmune diseases in MRL/lpr mice using a new strategy for allogeneic bone marrow transplantation. *Blood.* 2000;95(5):1862–1868.

[39] Simmons PJ, Przepiorka D, Thomas ED, Torok-Storb B. Host origin of marrow stromal cells following allogeneic bone marrow transplantation. *Nature.* 1987;328(6129):429–432.

[40] Laver J, Jhanwar SC, O'Reilly RJ, Castro-Malaspina H. Host origin of the human hematopoietic microenvironment following allogeneic bone marrow transplantation. *Blood.* 1987;70(6):1966–1968.

[41] Santucci MA, Trabetti E, Martinelli G, et al. Host origin of bone marrow fibroblasts following allogeneic bone marrow transplantation for chronic myeloid leukemia. *Bone Marrow Transplant.* 1992;10(3):255–259.

[42] Gordon MY. The origin of stromal cells in patients treated by bone marrow transplantation. *Bone Marrow Transplant.* 1988;3(4):247–251.

[43] Agematsu K, Nakahori Y. Recipient origin of bone marrow-derived fibroblastic stromal cells during all periods following bone marrow transplantation in humans. *Br J Haematol.* 1991;79(3):359–365.

[44] Cilloni D, Carlo-Stella C, Falzetti F, et al. Limited engraftment capacity of bone marrow-derived mesenchymal cells following T-cell-depleted hematopoietic stem cell transplantation. *Blood.* 2000;96(10):3637–3643.

[45] Keating A, Singer JW, Killen PD, et al. Donor origin of the *in vitro* haematopoietic microenvironment after marrow transplantation in man. *Nature.* 1982;298(5871):280–283.

[46] Awaya N, Rupert K, Bryant E, Torok-Storb B. Failure of adult marrow-derived stem cells to generate marrow stroma after successful hematopoietic stem cell transplantation. *Exp Hematol.* 2002;30(8):937–942.

[47] Koc ON, Peters C, Aubourg P, et al. Bone marrow-derived mesenchymal stem cells remain host-derived despite successful hematopoietic engraftment after allogeneic transplantation in patients with lysosomal and peroxisomal storage diseases. *Exp Hematol.* 1999;27(11):1675–1681.

[48] Horwitz EM, Prockop DJ, Fitzpatrick LA, et al. Transplantability and therapeutic effects of bone marrow-derived mesenchymal cells in children with osteogenesis imperfecta. *Nat Med.* 1999;5(3):309–313. See comments.

[49] Horwitz EM, Prockop DJ, Gordon PL, et al. Clinical responses to bone marrow transplantation in children with severe osteogenesis imperfecta. *Blood.* 2001;97(5):1227–1231.

[50] Koc ON, Lazarus HM. Mesenchymal stem cells: heading into the clinic. *Bone Marrow Transplant.* 2001;27(3):235–239.

[51] Lee OK, Kuo TK, Chen W-M, Lee K-D, Hsieh S-L, Chen T-H. Isolation of multi-potent mesenchymal stem cells from umbilical cord blood. *Blood.* 2003;5:1670.

[52] Lazarus HM, Haynesworth SE, Gerson SL, Caplan AI. Human bone marrow-derived mesenchymal (stromal) progenitor cells (MPCs) cannot be recovered from peripheral blood progenitor cell collections. *J Hematother.* 1997;6(5):447–455.

[53] Weimann JM, Johansson CB, Trejo A, Blau HM. Stable reprogrammed heterokaryons form spontaneously in Purkinje neurons after bone marrow transplant. *Nat Cell Biol.* 2003;5(11):959–966.

[54] Ying QL, Nichols J, Evans EP, Smith AG. Changing potency by spontaneous fusion. *Nature.* 2002;416(6880):545–548.

[55] Vassilopoulos G, Wang PR, Russell DW. Transplanted bone marrow regenerates liver by cell fusion. *Nature.* 2003;422(6934):901–904.

[56] Terada N, Hamazaki T, Oka M, et al. Bone marrow cells adopt the phenotype of other cells by spontaneous cell fusion. *Nature.* 2002;416(6880):542–545.

[57] Prockop DJ, Gregory CA, Spees JL. One strategy for cell and gene therapy: harnessing the power of adult stem cells to repair tissues. *PNAS.* 2003;100(90001):11917–11923.

[58] Wang X, Willenbring H, Akkari Y, et al. Cell fusion is the principal source of bone-marrow-derived hepatocytes. *Nature.* 2003;422(6934):897–901.

[59] Pereira RF, Halford KW, O'Hara MD, et al. Cultured adherent cells from marrow can serve as long-lasting precursor cells for bone, cartilage, and lung in irradiated mice. *Proc Natl Acad Sci USA.* 1995;92(11):4857–4861.

[60] Mosca JD, Hendricks JK, Buyaner D, et al. Mesenchymal stem cells as vehicles for gene delivery. *Clin Orthop.* 2000;379 (suppl):S71–S90.

[61] Devine SM, Cobbs C, Jennings M, Bartholomew A, Hoffman R. Mesenchymal stem cells distribute to a wide range of tissues following systemic infusion into nonhuman primates. *Blood.* 2003;101(8):2999–3001.

[62] Mahmud N, Pang W, Devine SM, et al. Unrelated allogeneic mesenchymal stem cell grafts are capable of engrafting in non-human primates. *Blood.* 2002;100(11):133a.

[63] Klyushnenkova E, Shustova V, Mosca J, Moseley A, McIntosh K. Human mesenchymal stem cells induce unresponsiveness in preactivated but not naive alloantigen specific T cells. *Exp Hematol.* 1999;27(7):122.

[64] Bartholomew A, Sturgeon C, Siatskas M, et al. Mesenchymal stem cells suppress lymphocyte proliferation *in vitro* and prolong skin graft survival *in vivo*. *Exp Hematol.* 2002;30(1):42–48.

[65] Ossevoort MA, Lorre K, Boon L, et al. Prolonged skin graft survival by administration of anti-CD80 monoclonal antibody with cyclosporin A. *J Immunother.* 1999;22(5):381–389.

[66] Goodman ER, Fiedor PS, Fein S, Athan E, Hardy MA. Fludarabine phosphate: a DNA synthesis inhibitor with potent immunosuppressive activity and minimal clinical toxicity. *The American Surgeon.* 1996;62(6):435–442.

[67] Di Nicola M, Carlo-Stella C, Magni M, et al. Human bone marrow stromal cells suppress T-lymphocyte proliferation induced by cellular or nonspecific mitogenic stimuli. *Blood.* 2002;99(10):3838–3843.

[68] Djouad F, Plence P, Bony C, et al. Immunosuppressive effect of mesenchymal stem cells favors tumor growth in allogeneic animals. *Blood.* 2003;102(10):3837–3844.

[69] Kim D-W, Chung Y-J, Kim T-G, Kim Y-L, Oh I-H. Cotransplantation of third-party mesenchymal stromal cells can alleviate one-donor predominance and increase engraftment from double cord transplantation. *Blood.* 2003;5:1601.

[70] Krampera M, Glennie S, Dyson J, et al. Bone marrow mesenchymal stem cells inhibit the response of naive and memory antigen-specific T cells to their cognate peptide. *Blood.* 2003;101(9):3722–3729.

[71] Tse WT, Pendleton JD, Beyer WM, Egalka MC, Guinan EC. Suppression of allogeneic T-cell proliferation by human marrow stromal cells: implications in transplantation. *Transplantation.* 2003;75(3):389–397.

[72] Rasmusson I, Ringden O, Sundberg B, Le Blanc K. Mesenchymal stem cells inhibit the formation of cytotoxic T lymphocytes, but not activated cytotoxic T lymphocytes or natural killer cells. *Transplantation.* 2003;76(8):1208–1213.

[73] Potian JA, Aviv H, Ponzio NM, Harrison JS, Rameshwar P. Veto-like activity of mesenchymal stem cells: Functional discrimination between cellular responses to alloantigens and recall antigens. *J Immunol*. 2003;171(7):3426–3434.

[74] Caplan AI, Bruder SP. Mesenchymal stem cells: building blocks for molecular medicine in the 21st century. *Trends Mol Med*. 2001;7(6):259–264.

[75] Keating A, Berkahn L, Filshie R. A phase I study of the transplantation of genetically marked autologous bone marrow stromal cells. *Hum Gene Ther*. 1998;9(4):591–600.

[76] Prockop DJ. Marrow stromal cells as stem cells for nonhematopoietic tissues. *Science*. 1997;276(5309):71–74.

[77] Schwarz EJ, Alexander GM, Prockop DJ, Azizi SA. Multipotential marrow stromal cells transduced to produce L-DOPA: engraftment in a rat model of Parkinson disease. *Hum Gene Ther*. 1999;10(15):2539–2549.

[78] Allay JA, Dennis JE, Haynesworth SE, et al. LacZ and interleukin-3 expression *in vivo* after retroviral transduction of marrow-derived human osteogenic mesenchymal progenitors. *Hum Gene Ther*. 1997;8(12):1417–1427.

[79] Hurwitz DR, Kirchgesser M, Merrill W, et al. Systemic delivery of human growth hormone or human factor IX in dogs by reintroduced genetically modified autologous bone marrow stromal cells. *Hum Gene Ther*. 1997;8(2):137–156.

[80] Wang G, Lee K, Liu L, et al. Human erythropoietin gene delivery using adult mesenchymal stem cells can prevent drug induced anemia in NOD/SCID mouse model. *Blood*. 1999;94:398a.

[81] Bartholomew A, Sturgeon C, Siatskas M, et al. Genetically modified mesenchymal stem cells can effectively deliver bioactive erythropoietin in the baboon. *Blood*. 1999;94:378a.

[82] Dunbar CE, Kohn DB, Schiffmann R, et al. Retroviral transfer of the glucocerebrosidase gene into CD34+ cells from patients with Gaucher disease: *in vivo* detection of transduced cells without myeloablation. *Hum Gene Ther*. 1998;9(17):2629–2640.

[83] Huhn RD, Tisdale JF, Agricola B, Metzger ME, Donahue RE, Dunbar CE. Retroviral marking and transplantation of rhesus hematopoietic cells by nonmyeloablative conditioning. *Hum Gene Ther*. 1999;10(11):1783–1790.

[84] Jin HK, Carter JE, Huntley GW, Schuchman EH. Intracerebral transplantation of mesenchymal stem cells into acid sphingomyelinase-deficient mice delays the onset of neurological abnormalities and extends their life span. *J Clin Invest*. 2002;109(9):1183–1191.

[85] Koc ON, Day J, Nieder M, Gerson SL, Lazarus HM, Krivit W. Allogeneic mesenchymal stem cell infusion for treatment of metachromatic leukodystrophy (MLD) and Hurler syndrome (MPS-IH). *Bone Marrow Transplant*. 2002;30(4):215–222.

[86] Mangi AA, Noiseux N, Kong D, et al. Mesenchymal stem cells modified with Akt prevent remodeling and restore performance of infarcted hearts. *Nat Med*. 2003;9(9):1195–201.

[87] Koc ON, Gerson SL. Akt helps stem cells heal the heart. *Nat Med*. 2003;9(9):1109–1110.

[88] Di Nicola M, Carlo-Stella C, Magni M, et al. Human bone marrow stromal cells suppress T-lymphocyte proliferation induced by cellular or nonspecific mitogenic stimuli. *Blood*. 2002;99(10):3838–3843.

CHAPTER 6

ENGINEERING OF HUMAN ADIPOSE-DERIVED MESENCHYMAL STEM-LIKE CELLS

J. K. FRASER, M. ZHU, B. STREM AND M. H. HEDRICK

Macropore Biosurgery Inc., San Diego, CA

1. INTRODUCTION

White adipose tissue is unique in being the only tissue which can dramatically change mass in the adult. Thus, while the normal range for adipose tissue mass expressed as a percentage of total body weight is 14–28% for females and 9–18% for males, athletes performing at the elite levels in their sport can have levels as low as 2–3%, while persons with obesity can have levels as high as 60–70% [1]. A positive energy balance generally results in increased size of adipose depots as a result of expansion of adipocyte volume which generally precedes, perhaps even triggers [2], increased cell number [3–5]. Proliferating cells include preadipocytes and adipose stem cells located in the adipose stromal vascular fraction [6–9]. Recent work has demonstrated that adipose tissue contains a population of cells that has the capacity to differentiate beyond the adipocytic lineage into, and perhaps beyond, other mesodermal tissues [10–15]. In this chapter we will review the biology of adipose tissue-derived mesenchymal stem cell-like cells, compare and contrast them with their counterparts from marrow, and briefly examine their therapeutic potential.

2. BACKGROUND

Marrow-derived mesenchymal stem cells, also referred to as marrow stromal cells and referred to collectively here as MSC, are well reviewed elsewhere in this book. Adipose-derived MSC-like stem cells (herein referred to as ADSC) are extracted by enzymatic digestion of adipose followed by removal of lipid-laden adipocytes and concentration of the non-buoyant cell fraction. This generates a heterogeneous population containing ADSC along with microvascular endothelial cells and smooth muscle cells [12]. When placed in culture under conditions supportive of MSC growth, a more homogeneous population emerges; a population which shares many properties of MSC including their extensive proliferative potential and the ability to undergo multilineage mesenchymal

J.A. Nolta (ed.), Genetic Engineering of Mesenchymal Stem Cells, 111–125.

differentiation. However, despite their many similarities, one notable feature immediately distinguishes ADSC from MSC; while proliferation and differentiation of MSC are exquisitely sensitive to differences between serum lots, it is rare to find a batch of serum that causes substantially reduced growth of ADSC [12]. This may be due to the greater frequency of stem cells in the adipocyte-depleted fraction.

A number of studies have used clonogenic assays to quantify MSC in marrow [16–19]. In these assays cells are plated at approximately $100,000/cm^2$ and grown for 2–3 weeks after which colonies of more than 50 cells are enumerated. Using these assays the number of MSC in bone marrow is generally found to be approximately 1 in 25,000 to 1 in 100,0000 [16–19] although many authors have found that frequency is influenced by factors such as age, gender, presence of osteoporosis, and prior exposure to high dose chemotherapy or radiation [16, 20–22]. Our preliminary, unpublished data suggest that the average frequency of such cells in processed lipoaspirate obtained from 56 donors (median age 49) is approximately 2% of nucleated cells although this is influenced by age, site from which the tissue was obtained, and donor body mass index. This is consistent with the observation that most investigators working with ADSC plate the fresh adipose-derived cells (processed lipoaspirate) at an initial density of $3,500$ cells/cm^2, substantially less than that generally used for fresh marrow cells.

The significance of this to tissue engineering with stem cells is clear. Donor site morbidity limits the amount of marrow that can be obtained and thereby extends the time in culture required to generate a therapeutic cell dose. Thus, the volume of human marrow taken under local anesthesia is generally limited to no more than 40 ml and yields approximately 1.2×10^9 nucleated cells [23]. Obtaining a larger volume necessitates use of general anesthesia, post-harvest morbidity [24, 25], and increasing dilution with stem cell-free blood [23]. At the stem cell frequency cited above [16, 17] this will contain approximately 1.2×10^5 MSC. By contrast, a typical harvest of adipose under local anesthesia can easily exceed 200 ml and yield approximately 2×10^8 nucleated cells per 100 ml of lipoaspirate [26] which, at 2% frequency, provides 4×10^6 stem cells per 100 ml lipoaspirate; a differential of approximately 30-fold more than that present in 40 ml of marrow.

2.1. Cell surface phenotype of ADSC

To date, three studies have examined the cell surface phenotype of ADSC; two examined a large panel of markers [13, 27], while the third looked at a smaller panel and compared MSC and ADSC obtained from the same human donor and prepared under identical conditions. Overall the cell surface phenotype of ADSC is very similar to that of MSC. Both populations express CD105 (endoglin), SH3 [28], Stro-1 [29], CD9, CD29, CD44, CD54, CD55, and CD90. ADSC and MSC are also CD45 and CD34-negative (or very low). However, it should be noted that there is some variation between donors in expression of surface markers [30, 31], a phenomenon that has also been recognized for MSC in differences in surface phenotype, proliferation, and growth requirements in different rodent strains [32, 33].

Some consistent differences in surface antigen expression between ADSC and MSC have been described. CD49d (α4 integrin) is a cell surface molecule, which, in combination with CD29, forms a heterodimer referred to as VLA-4. The ligand for VLA-4 is a molecule referred to as VCAM-1 (Vascular Cell Adhesion Molecule-1 also known as CD106) and the interaction between these molecules has been shown to play an important role in hematopoietic stem cell mobilization from, and homing to, the marrow [34–36]. There is general agreement in the literature that MSC express VCAM-1 [36–38] and that they do not express CD49d/VLA-4 [38, 39]. These observations are confirmed by our studies which also demonstrate the reciprocal pattern of expression of these cognate molecules in that ADSC express CD49d but not VCAM-1 [13, 31].

3. MULTILINEAGE DIFFERENTIATION CAPACITY

We and others have demonstrated the ability of adipose tissue-derived cells to undergo differentiation along classical mesenchymal lineages with most studies focusing on adipogenesis, chondrogenesis, osteogenesis, and myogenesis [11–13, 15, 30, 40–46]. Our data using single cell-derived clonal populations demonstrates that at least part of this plasticity resides in a population of multipotential cells [13]. In most respects these data demonstrate a set of functional properties that is very similar to that of MSC; however, some differences have been observed.

3.1. Adipose

Given the origin of ADSC it is not surprising that, when cultured in adipogenic medium, ADSC express several adipocytic genes including lipoprotein lipase, aP2, PPARγ2, leptin, Glut4, and develop lipid-laden intracellular vacuoles, the definitive marker of adipogenesis [12, 13, 15, 47]. Despite certain donor-to-donor qualitative differences in adipogenic potential [47] the pattern of expression of these genes appears to be very similar, if not identical, to that observed for adipogenic differentiation in MSC [32, 38, 39]. The *in vivo* capacity of these cells to differentiate into cells of the adipocytic lineage has also been demonstrated in studies involving implantation of cell-seeded natural (collagen and hyaluronic) [48–50] or synthetic bioresorbable (polylactic acid, polyglycolic acid) [51, 52] scaffolds. It is important to note that these studies generally agree that robust ectopic *in vivo* adipogenesis requires *in vitro* pre-differentiation of adipose tissue-derived cells prior to implantation. This requirement may be eliminated by co-implantation of cells with a source of adipogenic stimuli. Thus, Yuksel and colleagues have demonstrated ectopic adipogenesis at the site of implantation of microbeads containing insulin and insulin-like growth factor 1 [53]. This suggests the recruitment of adipocytic stem and progenitor cells to the site of implantation, but it is not clear if this is a local recruitment or derives from distal compartments of progenitors in fat and/or marrow.

3.2. Bone

Culture of ADSC and MSC in osteogenic medium leads to a well-defined pattern of expression of genes associated with the deposition and subsequent mineralization of a collagen matrix [12, 13, 15, 30, 43, 54–56]. Thus, we and others have demonstrated expression of the osteogenic master-switch CBFA-1, bone sialoprotein, alkaline phosphatase, collagen I, osteonectin, osteocalcin, and osteopontin as well as *in vitro* mineral deposition by ADSC. The osteogenic capacity of MSC and ADSC has been compared by Dragoo et al. [40]. In this study the authors demonstrated that ADSC treated with BMP-2 produced more bone precursors (as assessed by production of alkaline phosphatase and calcified extracellular matrix) than MSC ($p \leq 0.001$). The data further show that ADSC transduced with an adenovirus encoding BMP2 produce faster onset of calcified extracellular matrix than transduced MSC. Our own data show some subtle differences in osteogenesis in that ADSC exhibited significantly elevated alkaline phosphatase levels compared to MSC controls at 3 weeks of induction [13]. However, despite the lower enzyme activity compared with ADSC, induced MSC were associated with significantly more matrix calcification, compared with induced ADSC. It should be noted that these differences may reflect cellular heterogeneity of the populations used to initiate the cultures rather than inherent differences in the stem cells themselves. Thus, bone marrow might reasonably be expected to have a greater diversity of osteochondral progenitor cells than adipose [57] and that the superimposition of these functional properties of these more mature cells onto those of the stem cells gives rise to differences in osteochondral behavior of the two populations. The hypothesis that marrow-derived cultures might have a broader diversity of osteochondral progenitors, specifically of more mature cells, is supported by the observation that low level expression of osteopontin and osteocalcin has been detected in uninduced MSC [13, 44]. It should also be noted that Hicok et al. have demonstrated *in vivo* generation of osteoid by human ADSC seeded into ceramic cubes (HA-TCP) implanted in immunodeficient mice [58]. Similarly, Cowan et al. have shown that autologous ADSC seeded onto an apatite-coated resorbable scaffold heal a critical size calvarial defect in mice [59]. Another group has obtained robust *in vivo* production of bone by ADSC engineered to express BMP-2 using an adenovirus vector [41].

3.3. Cartilage

Culture of ADSC and MSC in high density micromass cultures results in generation of cellular nodules that are rich in sulfated proteoglycans (Alcian Blue-positivity) and which express both juvenile and mature splice variants of Collagen II, as well as collagen VI, aggrecan, PRELP, and other chondrocytic markers [12, 13, 15, 38, 41, 42, 44, 56]. In our studies using MSC and ADSC obtained from the same donor and cultured under identical conditions, we found that ADSC had considerably greater chondrogenic capacity than MSC [30]. In contrast, work by Winter et al. demonstrated that though these two cell types exhibited essentially identical chondrogenesis in 2-dimensional cultures, MSC showed more robust chondrogenesis in 3-dimensional cultures [44]. Thus, in 3-dimensional culture MSC-derived cells expressed a greater number of genes associated with mature cartilage than ADSC-derived cultures. This difference suggests that ADSC may be less useful for engineering cartilage. However, a recent study has demonstrated *in vivo* development of mature cartilage from ADSC in a rabbit osteochondral defect model [46]. The authors noted that the repair induced by ADSC was superior to that derived from osteochondral autografts. Specifically, the Pineda score, a composite score assessing four different parameters of cartilage repair, was greater for ADSC-derived grafts than for osteochondral grafts at each time point examined. ADSC grafts also showed superior performance in creep indentation biomechanical testing performed at 24 weeks. However, it should be noted that performance at 24 weeks was still inferior to intact cartilage. Nonetheless, these data indicate that the deficit observed by Winter et al. may be a culture artifact rather than an inherent limitation in the osteochondral capacity of ADSC. This is supported by studies in which ADSC seeded onto alginate discs and implanted into immunodeficient mice exhibited prolonged (12 week) synthesis of cartilage matrix molecules including collagen II, collagen VI, and aggrecan [42].

3.4. Intervertebral disc

Cell and gene therapy for the intervertebral disc, and in particular, the central nucleus pulposus (NP) has the potential to address the increasing problem of degenerative disc disease (DDD) [60–63]. The ability of MSC to generate cells capable of producing a proteoglycan-rich matrix has led to the proposal for their use in this setting. Thus, Sakai et al. have used a rabbit disc injury model in which they demonstrated the ability of autologous Ad-lacZ-marked MSC to survive in the essentially avascular environment of the NP. Survival was associated with proteoglycan accumulation and preservation of disc structure when compared to discs treated with carrier alone [64]. While this study was limited by its short duration (4 weeks) and by the clinical relevance of using skeletally immature animals, it suggests value for stem cell therapy in the disc. Others have used Ad-luciferase transduction to permit *in vivo* bioluminescent tracking of ADSC in rat intervertebral discs [65]. In this study Ad-luciferase-transduced ADSC were suspended in alginate and implanted into lumbar spine intervertebral discs of Sprague-Dawley rats. Using non-invasive, real-time imaging, transduced cells were observed at the site of injection for the duration of the study (14d). This suggests that, like MSC, ADSC are capable of survival and continued metabolic activity in the relatively hostile environment of the intervertebral disc. Other investigators have applied gene transfer to NP-derived cells using genes directed at matrix remodeling [66], growth factors [67], and a chondrogenic transcription factor [68]. It is likely that success in these models may lead to application with MSC or ADSC.

3.5. Skeletal muscle

Culture of ADSC and MSC in the presence of dexamethasone, hydrocortisone, and/or 5 azacytidine results in a time-dependent pattern of expression of muscle-related genes that is consistent with normal myogenesis defined by early expression of key master regulatory factors MyoD1 and myf5, myf6, and myogenin followed by later expression of myosin heavy chain [11, 13, 69]. This process is associated with characteristic changes in cell morphology including generation of long, multinucleate, MyoD1-positive cells

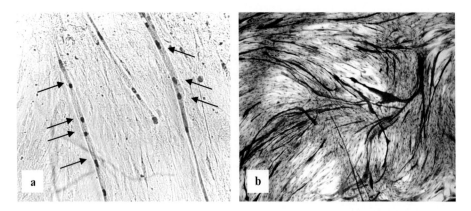

Figure 6.1. Expression of skeletal muscle-specific genes by ADSC. (A) MyoD immunostaining (arrows) (40x obj)1; (B) myosin heavy chain immunostaining (10x obj).

(Figure 6.1.A) early in culture and bundles of myosin heavy chain-positive myofibrillar structures (Figure 6.1.B) appearing after 2 weeks.

3.6. Cardiac muscle

While *in vitro* transdifferentiation of skeletal muscle cells into cardiac myocytes has been reported [70] this phenomenon has not been repeated by other laboratories in a number of preclinical and clinical studies [71–75]. However, in addition to the skeletal muscle differentiation described above both ADSC [76–78] and MSC [79–82] have been shown to be capable of *in vitro* differentiation into cardiac myocytes. The most compelling data was obtained by Planat-Bernard et al. in which fresh adipose-derived cells were plated into semisolid culture. After 3 weeks colonies of spontaneously beating cells were observed. These cells exhibited several molecular, electrophysiologic, and pharmacologic properties of cardiac myocytes [77].

Research with MSC has been extended to *in vivo* studies which have demonstrated homing of MSC to the site of injury within the heart [83] and the ability of both unmodified [84–86] and transduced [87] MSC to contribute to cardiac repair in clinically substantial fashion. Most notably, a recent paper in Nature Medicine demonstrated that transplantation of syngeneic MSC that had been transduced with the survival gene Akt resulted in considerable reduction in infarct size, scaring, and improvements in myocardial function following experimental occlusion of the coronary artery in rats [87]. Implantation of 5 million transduced cells resulted in normalization of systolic and diastolic cardiac function and regeneration of 80–90% of lost myocardial volume. We have examined *in vivo* myocardial differentiation of ADSC in studies in which fresh adipose-derived cells obtained from lacZ expressing transgenic mice (Rosa26) were injected into the intraventricular chamber of mice immediately following induction of cardiac cryoinjury. Two weeks following injury, lacZ-positive cells co-expressing cardiac markers were detected in the infarcted region of treated mice (Figure 6.2). Further,

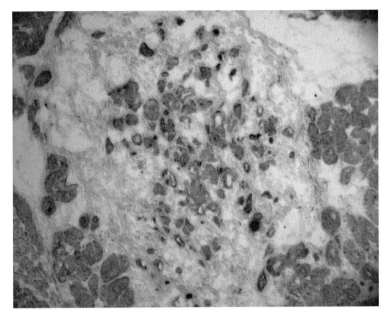

Figure 6.2. Expression of cardiac-specific myosin heavy chain (MHC) by donor cells located in the infarct region. Staining shows infarct (clear area) surrounded by healthy MHC-positive myocardium. In the center of the infarct are MHC-positive cells, which exhibit nuclear staining for the lacZ transgene product indicating donor cell origin.

at the recent annual meeting of the European Society of Cardiology, Valina et al. presented data showing that intracoronary infusion of 2×10^6 ADSC 15 minutes following a 3-hour coronary occlusion in farm swine resulted in preservation of myocardium and improvement in global cardiac function.

3.7. Hematopoietic support cell (marrow stroma)

The ability to support hematopoiesis is another property of MSC that may be important in clinical applications. Indeed, co-infusion of MSC with grafts containing hematopoietic stem cells has been shown to enhance the rate of hematopoietic engraftment in human clinical studies [88]. Hematopoietic support is also important in transduction of CD34-positive hematopoietic stem and progenitor cells [89–91]. While no study to date has specifically examined the ability of ADSC to support hematopoiesis, a recent study has claimed to demonstrate that adipose tissue contains a population of cells with hematopoietic stem cell activity; that is, a population of cells capable of rescuing lethally irradiated animals [92]. Thus, intraperitoneal transplant of 10^7 fresh adipose or marrow cells was associated with 40% survival following 10Gy irradiation. Recovery of platelet and white blood cells counts was more rapid with marrow than for adipose tissue cells (8 weeks vs 10 weeks for return to normal levels). However, adipose-derived cell transplant resulted in a very low level of hematopoietic chimerism (1.7% marrow hematopoietic progenitor cells of donor origin). This suggests that the survival advantage conferred by adipose-derived cell transplant is due to enhancement of the recovery of *endogenous* hematopoietic stem cells from the otherwise lethal irradiation in a manner that is generally consistent with the human co-infusion study cited above [88].

3.8. Neuronal

In vitro differentiation along the neuronal lineages has also been demonstrated for both ADSC [10, 13, 14] and MSC [93, 94]. Thus, treatment of rat and human MSC or ADSC with beta-mercaptoethanol results in rapid transition of cells to a neuronal morphology (a condensed cell body with multiple neuron-like outgrowths), and expression of neuronal markers including nestin, neuron-specific enolase (NSE), and neuron-specific protein (NeuN) all of which are early markers of the neuronal lineage. Similar results are seen with alternate inductive conditions such as isobutylmethylxanthine (IBMX) and dibutyryl cAMP or forskolin and butylated hydroxyanisole. We have also detected expression of trkA (a receptor for NGF) and the presence of voltage gated potassium channels. However, to date, detection of neuronal markers in *in vitro* differentiated ADSC and MSC has been restricted to these early genes; no expression of markers characteristic of mature neurons, oligodendrocytes, or astrocytes has been described. This may suggest that the expression of these markers is the result of disordered gene expression resulting from the toxic inductive stimulus or that the induction is unmasking an inherent neuronal potential that is only partially supported by the culture conditions.

The latter interpretation is supported by *in vivo* studies in which ADSC and MSC have been implanted into the CNS of experimental animals. Zhao et al. [95] have demonstrated that implantation of MSC into the cortex of rats following ischemic injury resulted in significantly improved performance in a limb placement test and that the implanted cells had undergone an *in vivo* change in marker expression consistent with differentiation along astrocytic, oligodendrocytic, and neural lineages. However, there was no evidence for incorporation of these cells into the cerebral architecture. Therefore, it is possible that the observed functional improvement was due to an indirect mechanism, for example, paracrine expression of angiogenic and/or anti-apoptotic factors by the implanted cells would promote survival of functionally compromised but viable host tissue [96, 97]. Other studies have also demonstrated engraftment of MSC into the region in the absence of injury albeit with modest evidence of differentiation [98, 99]. Nonetheless, these data provide substantial support for a potential role of MSC in direct or indirect (gene-modified) therapy for the CNS [100, 101]. ADSC have also been applied in the setting of experimental stroke. Thus, Kang et al. directly implanted ADSC into the brain of rats following 90 minutes of middle cerebral artery occlusion [102]. In some studies the ADSC were transduced with either the lacZ gene or the gene encoding brain-derived neurotrophic factor (BDNF) as both a marker and a potentially therapeutic agent. Marked cells were seen throughout the infarct area 14 days after implantation and a fraction of these cells co-expressed MAP2 (4% of marked cells) or GFAP (9% of marked cells) suggesting a degree of neuronal differentiation. No data were presented with regard to potential fusion between donor and recipient cells. As with the MSC studies, ADSC-treated animals showed significant improvement in neurologic testing with the animals receiving BDNF-transduced cells exhibiting significantly better recovery of function than those treated with unmodified ADSC. The same group has also demonstrated that co-culture of ADSC and neural stem cells (NSC) decreases the proliferation but increases the neuronal differentiation of the NSCs [103].

3.9. Endothelium and hematopoiesis

Sampaolesi et al. have recently described an apparently novel vessel-associated stem cell with dual mesenchymal and angioblast capacity [104, 105]. This population was originally described in the fetal dorsal aorta [104] but has also been detected, albeit in much smaller numbers in the post-natal vasculature. These cells, termed mesoangioblasts share many properties with ADSC including extensive proliferative capacity and multilineage, mesenchymal differentiation. They also possess the ability to differentiate into both hematopoietic and endothelial lineages. A review of the literature provides evidence suggesting the presence of cells with similar capacity within the adipose compartment. For example, Cousin et al. have demonstrated the ability of adipose tissue-derived cells to support hematopoietic recovery in mice following lethal irradiation [92]. However, in this study, nested PCR demonstrated donor origin of only $1.7 \pm 0.5\%$ of bone-derived CFU-GM colonies (colonies of granulocytes and macrophages derived from a hematopoietic progenitor) and $5.2 \pm 3.2\%$ of splenic CFU-GM. Thus, while there may be some hematopoietic stem or progenitor cells within adipose tissue, it appears that ADSC-mediated recovery of *host* hematopoiesis is the primary mechanism by which recovery from otherwise lethal irradiation was mediated.

Miranville et al. have presented data supporting the presence of cells within adipose tissue that differentiated into endothelium [106]. Thus, $CD34^+/CD31^-$ cells within adipose were shown to be capable of *in vitro* differentiation into cells that expressed both CD31 and von Willebrand factor, both markers of mature endothelium. Most importantly, the authors demonstrated the ability of these cells to improve blood flow and capillary density in a NOD-SCID mouse model of hind limb ischemia. These data are confirmed by another study showing that delivery of ADSC to immunodeficient animals following induction of severe hind limb ischemia results in accelerated restoration of perfusion [107]. As an interesting side note Miranville et al. also reported that approximately 18% of $CD34^+/CD31^-$ cells co-expressed ABCG2, a protein associated with the side population (SP) stem cell phenotype; overall approximately 4% of all non-buoyant adipose-tissue-derived cells express ABCG2 [106]. It should be noted that there, at present, are no data addressing the question of whether or not the endothelial and mesenchymal differentiation capacity within adipose tissue reside within the same cells.

4. TRANSDUCTION OF ADSC

As noted above a number of investigators have transduced ADSC in order to facilitate tracking or to engender a therapeutic effect. Thus, Leo et al. used Ad-CMV-luciferase to allow non-invasive, real-time tracking of ADSC in rat spine [65]. Similarly, Dragoo et al. [40] infected both MSC and ADSC with E1A-deleted type 5 adenovirus constructs containing the BMP-2 (bone morphogenic protein-2) gene or the bacterial beta-galactosidase (lacZ) gene. LacZ gene transduction efficiency was 35% for MSC and 55% for ADSC. Ad BMP2 infection resulted in levels of expression of BMP-2 protein that were threefold higher than those derived from MSC. Ad-BMP-2 infected ADSC exhibited *in vitro* osteoblastic differentiation in the absence of exogenous osteogenic factors. They also exhibited robust ectopic *in vivo* production of bone when cells were implanted into a collagen sponge within the subcutaneous space [40]. Given the success of unmodified MSC in treatment of osteogenesis imperfecta [108, 109] these data support the potential for transplant of allogeneic or gene-modified ADSC for genetic disorders of the skeletal system.

Kang et al. have also used an E1A-deleted type 5 adenovirus to infect ADSC. As above these studies employed transduction of a tracking gene (lacZ) and a potentially therapeutic gene Brain-Derived Neurotrophic Factor (BDNF) achieving 100% and 94% transduction efficiency respectively. Transduced cells were implanted into areas of the brain that had undergone transient (90 minute) ischemia/reperfusion injury. Donor cells capable of continued expression of the transgene were retained to 30 days, the longest timepoint examined in this study.

Finally, our group has published results of a study comparing infection of ADSC with oncoretroviral, and lentiviral vectors [110]. The primary lentivirus used was the VSV-G pseudotyped HIV-1 vector SIN18-Rh-MLV-E (VSV). Infection by the VSV-G pseudotyped MuLV oncoretrovirus SRαL-EGFP and a second lentivirus construct, RRL-PGK-EGFP-SIN18, was also used. Direct comparison of infection of the lentiviruses and the oncoretrovirus was possible due to common envelope protein and the similarity of the transcription level driven in transduced cells. Thus, we were able to infect ADSC with the same number of EGFP transduction units (virus preparations standardized to drive the same level of GFP expression in control cells) and determine efficiency by flow cytometry 3 days and 1 week after infection. The lentiviral contructs resulted in four- to tenfold higher expression than the retroviral vector. The percentage of transduced cells was not high (10–15%) but remained stable in culture over 100 days. Moreover, using a lentiviral vector with the cytomegalovirus promoter resulted in a transduction efficiency of >90% at a MOI of 14. Studies using lentiviral-infected cells (RRL-PGK-EGFP-SIN18; MOI 59) in which transduction efficiency was 98% at day 3 and >95% at day 100 allowed examination of gene expression during *in vitro* differentiation. Retention of marker gene (EGFP) expression was observed following both adipogenic and osteogenic differentiation.

In the light of the potential similarity of ADSC and mesoangioblasts discussed above, it is noteworthy that Sampaolesi et al. have used lentivirus (hPGK-GFP-α-SG vector) to infect mesoangioblasts at an MOI of 200, achieving a transduction efficiency of over 90% [105]. Tranduced cells were delivered by three intra-arterial doses of 5×10^5 cells into an α-SG null mouse model of muscular dystrophy. Three months later treated mice exhibited significant restoration of muscle function and motility. Improvement in function was also demonstrated by treatment with syngeneic wild-type mesoangioblasts.

5. SUMMARY

In summary, adipose contains a population of cells that has extensive self-renewal capacity and the ability to differentiate along multiple lineages. The cells possessing this activity can be obtained in large numbers at high frequency from a tissue source that can be extracted in large quantities with minimal morbidity. These cells can be infected by adenoviral, oncoretroviral, and lentiviral vectors with moderate to high efficiency. Thus, adipose tissue appears to represent a potential clinically useful source of cells for cellular therapy and gene transfer applications.

REFERENCES

[1] DiGirolamo M, Fine JB. Obesity. In: Branch WJ, Alexander R, Schlant R, Hurst J, eds. *Cardiology in Primary Care*. New York: McGraw-Hill; 2000:265–278.

[2] Bjorntorp P, Karlsson M, Pettersson P. Expansion of adipose tissue storage capacity at different ages in rats. *Metabolism*. 1982;31:366–373.

[3] Faust IM, Johnson PR, Stern JS, Hirsch J. Diet-induced adipocyte number increase in adult rats: a new model of obesity. *Am J Physiol*. 1978;235:E279–E286.

[4] Faust IM, Johnson PR, Hirsch J. Adipose tissue regeneration following lipectomy. *Science*. 1977;197:391–393.

[5] Miller WH Jr, Faust IM, Hirsch J. Demonstration of de novo production of adipocytes in adult rats by biochemical and radioautographic techniques. *J Lipid Res*. 1984;25:336–347.

[6] Deslex S, Negrel R, Vannier C, Etienne J, Ailhaud G. Differentiation of human adipocyte precursors in a chemically defined serum-free medium. *Int J Obes*. 1987;11:19–27.

[7] Deslex S, Negrel R, Ailhaud G. Development of a chemically defined serum-free medium for differentiation of rat adipose precursor cells. *Exp Cell Res*. 1987;168:15–30.

[8] Ailhaud G, Grimaldi P, Negrel R. Cellular and molecular aspects of adipose tissue development. *Annu Rev Nutr*. 1992;12:207–233.

[9] Pettersson P, Van R, Karlsson M, Bjorntorp P. Adipocyte precursor cells in obese and nonobese humans. *Metabolism*. 1985;34:808–812.

[10] Ashjian PH, Elbarbary AS, Edmonds B, et al. *In vitro* differentiation of human processed lipoaspirate cells into early neural progenitors. *Plast Reconstr Surg*. 2003;111:1922–1931.

[11] Mizuno H, Zuk PA, Zhu M, et al. Myogenic differentiation by human processed lipoaspirate cells. *Plast Reconstr Surg*. 2002;109:199–209.

[12] Zuk PA, Zhu M, Mizuno H, et al. Multilineage cells from human adipose tissue: implications for cell-based therapies. *Tissue Eng*. 2001;7:211–228.

[13] Zuk PA, Zhu M, Ashjian P, et al. Human adipose tissue is a source of multipotent stem cells. *Mol Biol Cell*. 2002;13:4279–4295.

[14] Safford KM, Hicok KC, Safford SD, et al. Neurogenic differentiation of murine and human adipose-derived stromal cells. *Biochem Biophys Res Commun*. 2002;294:371–379.

[15] Wickham MQ, Erickson GR, Gimble JM, Vail TP, Guilak F. Multipotent stromal cells derived from the infrapatellar fat pad of the knee. *Clin Orthop*. 2003;196–212.

[16] D'Ippolito G, Schiller PC, Ricordi C, Roos BA, Howard GA. Age-related osteogenic potential of mesenchymal stromal stem cells from human vertebral bone marrow. *J Bone Miner Res*. 1999;14:1115–1122.

[17] Muschler GF, Nitto H, Boehm CA, Easley KA. Age- and gender-related changes in the cellularity of human bone marrow and the prevalence of osteoblastic progenitors. *J Orthop Res*. 2001;19:117–125.

[18] Banfi A, Bianchi G, Galotto M, Cancedda R, Quarto R. Bone marrow stromal damage after chemo/radiotherapy: occurrence, consequences and possibilities of treatment. *Leuk Lymphoma*. 2001;42:863–870.

[19] Banfi A, Podesta M, Fazzuoli L, et al. High-dose chemotherapy shows a dose-dependent toxicity to bone marrow osteoprogenitors: a mechanism for post-bone marrow transplantation osteopenia. *Cancer*. 2001;92:2419–2428.

[20] Galotto M, Berisso G, Delfino L, et al. Stromal damage as consequence of high-dose chemo/radiotherapy in bone marrow transplant recipients. *Exp Hematol*. 1999;27:1460–1466.

[21] Oreffo RO, Bord S, Triffitt JT. Skeletal progenitor cells and ageing human populations. *Clin Sci (Lond)*. 1998;94:549–555.

[22] Stenderup K, Justesen J, Eriksen EF, Rattan SI, Kassem M. Number and proliferative capacity of osteogenic stem cells are maintained during aging and in patients with osteoporosis. *J Bone Miner Res*. 2001;16:1120–1129.

[23] Bacigalupo A, Tong J, Podesta M, et al. Bone marrow harvest for marrow transplantation: effect of multiple small (2 ml) or large (20 ml) aspirates. *Bone Marrow Transplant*. 1992;9:467–470.

[24] Auquier P, Macquart-Moulin G, Moatti JP, et al. Comparison of anxiety, pain and discomfort in two procedures of hematopoietic stem cell collection: leukacytapheresis and bone marrow harvest. *Bone Marrow Transplant*. 1995;16:541–547.

[25] Nishimori M, Yamada Y, Hoshi K, et al. Health-related quality of life of unrelated bone marrow donors in Japan. *Blood*. 2002;99:1995–2001.

[26] Aust L, Devlin B, Foster SJ, et al. Yield of human adipose-derived adult stem cells from liposuction aspirates. *Cytotherapy*. 2004;6:7–14.

[27] Gronthos S, Franklin DM, Leddy HA, et al. Surface protein characterization of human adipose tissue-derived stromal cells. *J Cell Physiol*. 2001;189:54–63.

[28] Haynesworth SE, Baber MA, Caplan AI. Cell surface antigens on human marrow-derived mesenchymal cells are detected by monoclonal antibodies. *Bone*. 1992;13:69–80.

[29] Dennis JE, Carbillet JP, Caplan A, Charbord P. The STRO-1+ marrow cell population is multipotential. *Cells Tissues Organs*. 2002;170:73–82.

[30] De Ugarte DA, Morizono K, Elbarbary A, et al. Comparison of multi-lineage cells from human adipose tissue and bone marrow. *Cells Tissues Organs*. 2003;174:101–109.

[31] De Ugarte DA, Alfonso Z, Zuk PA, et al. Differential expression of stem cell mobilization-associated molecules on multi-lineage cells from adipose tissue and bone marrow. *Immunol Lett*. 2003;89:267–270.

[32] Peister A, Mellad JA, Larson BL, et al. Adult stem cells from bone marrow (MSCs) isolated from different strains of inbred mice vary in surface epitopes, rates of proliferation, and differentiation potential. *Blood*. 2003.

[33] Phinney DG, Kopen G, Isaacson RL, Prockop DJ. Plastic adherent stromal cells from the bone marrow of commonly used strains of inbred mice: variations in yield, growth, and differentiation. *J Cell Biochem*. 1999;72:570–585.

[34] Oostendorp RA, Dormer P. VLA-4-mediated interactions between normal human hematopoietic progenitors and stromal cells. *Leuk Lymphoma*. 1997;24:423–435.

[35] Papayannopoulou T, Priestley GV, Nakamoto B. Anti-VLA4/VCAM-1-induced mobilization requires cooperative signaling through the kit/mkit ligand pathway. *Blood*. 1998;91:2231–2239.

[36] Simmons PJ, Masinovsky B, Longenecker BM, et al. Vascular cell adhesion molecule-1 expressed by bone marrow stromal cells mediates the binding of hematopoietic progenitor cells. *Blood*. 1992;80:388–395.

[37] Juneja HS, Schmalsteig FC, Lee S, Chen J. Vascular cell adhesion molecule-1 and VLA-4 are obligatory adhesion proteins in the heterotypic adherence between human leukemia/lymphoma cells and marrow stromal cells. *Exp Hematol*. 1993;21:444–450.

[38] Pittenger MF, Mackay AM, Beck SC, et al. Multilineage potential of adult human mesenchymal stem cells. *Science*. 1999;284:143–147.

[39] Conget PA, Minguell JJ. Phenotypical and functional properties of human bone marrow mesenchymal progenitor cells. *J Cell Physiol*. 1999;181:67–73.

[40] Dragoo JL, Choi JY, Lieberman JR, et al. Bone induction by BMP-2 transduced stem cells derived from human fat. *J Orthop Res*. 2003;21:622–629.

[41] Dragoo JL, Samimi B, Zhu M, et al. Tissue-engineered cartilage and bone using stem cells from human infrapatellar fat pads. *J Bone Joint Surg Br*. 2003;85:740–747.

[42] Erickson GR, Gimble JM, Franklin DM, et al. Chondrogenic potential of adipose tissue-derived stromal cells *in vitro* and *in vivo*. *Biochem Biophys Res Commun*. 2002;290:763–769.

[43] Halvorsen YD, Franklin D, Bond AL, et al. Extracellular matrix mineralization and osteoblast gene expression by human adipose tissue-derived stromal cells. *Tissue Eng*. 2001;7:729–741.

[44] Winter A, Breit S, Parsch D, et al. Cartilage-like gene expression in differentiated human stem cell spheroids: a comparison of bone marrow-derived and adipose tissue-derived stromal cells. *Arthritis Rheum.* 2003;48:418–429.

[45] Tholpady SS, Katz AJ, Ogle RC. Mesenchymal stem cells from rat visceral fat exhibit multipotential differentiation *in vitro. Anat Rec.* 2003;272A:398–402.

[46] Nathan S, Das DS, Thambyah A, et al. Cell-based therapy in the repair of osteochondral defects: a novel use for adipose tissue. *Tissue Eng.* 2003;9:733–744.

[47] Sen A, Lea-Currie YR, Sujkowska D, et al. Adipogenic potential of human adipose derived stromal cells from multiple donors is heterogeneous. *J Cell Biochem.* 2001;81:312–319.

[48] Halbleib M, Skurk T, de Luca C, von Heimburg D, Hauner H. Tissue engineering of white adipose tissue using hyaluronic acid-based scaffolds. I: *in vitro* differentiation of human adipocyte precursor cells on scaffolds. *Biomaterials.* 2003;24:3125–3132.

[49] von Heimburg D, Zachariah S, Heschel I, et al. Human preadipocytes seeded on freeze-dried collagen scaffolds investigated *in vitro* and *in vivo. Biomaterials.* 2001;22:429–438.

[50] von Heimburg D, Zachariah S, Low A, Pallua N. Influence of different biodegradable carriers on the *in vivo* behavior of human adipose precursor cells. *Plast Reconstr Surg.* 2001;108:411–420.

[51] Patrick CW Jr, Chauvin PB, Hobley J, Reece GP. Preadipocyte seeded PLGA scaffolds for adipose tissue engineering. *Tissue Eng.* 1999;5:139–151.

[52] Lee JA, Parrett BM, Conejero JA, et al. Biological alchemy: engineering bone and fat from fat-derived stem cells. *Ann Plast Surg.* 2003;50:610–617.

[53] Yuksel E, Weinfeld AB, Cleek R, et al. De novo adipose tissue generation through long-term, local delivery of insulin and insulin-like growth factor-1 by PLGA/PEG microspheres in an *in vivo* rat model: a novel concept and capability. *Plast Reconstr Surg.* 2000;105:1721–1729.

[54] Phinney DG, Kopen G, Righter W, et al. Donor variation in the growth properties and osteogenic potential of human marrow stromal cells. *J Cell Biochem.* 1999;75:424–436.

[55] Lennon DP, Haynesworth SE, Young RG, Dennis JE, Caplan AI. A chemically defined medium supports *in vitro* proliferation and maintains the osteochondral potential of rat marrow-derived mesenchymal stem cells. *Exp Cell Res.* 1995;219:211–222.

[56] Caplan AI. Mesenchymal stem cells. *J Orthop Res.* 1991;9:641–650.

[57] Solchaga LA, Cassiede P, Caplan AI. Different response to osteo-inductive agents in bone marrow- and periosteum-derived cell preparations. *Acta Orthop Scand.* 1998;69:426–432.

[58] Hicok KC, Du Laney TV, Zhou YS, et al. Human adipose-derived adult stem cells produce osteoid *in vivo. Tissue Eng.* 2004;10:371–380.

[59] Cowan CM, Shi YY, Aalami OO, et al. Adipose-derived adult stromal cells heal critical-size mouse calvarial defects. *Nat Biotechnol.* 2004;22:560–567.

[60] Ganey T, Libera J, Moos V, et al. Disc chondrocyte transplantation in a canine model: a treatment for degenerated or damaged intervertebral disc. *Spine.* 2003;28:2609–2620.

[61] Ganey TM, Meisel HJ. A potential role for cell-based therapeutics in the treatment of intervertebral disc herniation. *Eur Spine J.* 2002;11(suppl 2):S206–S214.

[62] Sato M, Asazuma T, Ishihara M, et al. An experimental study of the regeneration of the intervertebral disc with an allograft of cultured annulus fibrosus cells using a tissue-engineering method. *Spine.* 2003;28:548–553.

[63] Cassinelli EH, Hall RA, Kang JD. Biochemistry of intervertebral disc degeneration and the potential for gene therapy applications. *Spine J.* 2001;1:205–214.

[64] Sakai D, Mochida J, Yamamoto Y, et al. Transplantation of mesenchymal stem cells embedded in atelocollagen gel to the intervertebral disc: a potential therapeutic model for disc degeneration. *Biomaterials.* 2003;24:3531–3541.

[65] Leo BM, Li X, Balian G, Anderson DG. *In vivo* bioluminescent imaging of virus-mediated gene transfer and transduced cell transplantation in the intervertebral disc. *Spine.* 2004;29:838–844.

[66] Wallach CJ, Sobajima S, Watanabe Y, et al. Gene transfer of the catabolic inhibitor TIMP-1 increases measured proteoglycans in cells from degenerated human intervertebral discs. *Spine.* 2003;28:2331–2337.

[67] Tan J, Hu Y, Zheng H, Li S. Construction of recombinant adenoviral vector Ad-CMV-hTGFbeta1 for reversion of intervertebral disc degeneration by gene transfer. *Chin J Traumatol.* 2002;5:97–102.

[68] Paul R, Haydon RC, Cheng H, et al. Potential use of sox9 gene therapy for intervertebral degenerative disc disease. *Spine.* 2003;28:755–763.

[69] Wakitani S, Saito T, Caplan AI. Myogenic cells derived from rat bone marrow mesenchymal stem cells exposed to 5-azacytidine. *Muscle Nerve.* 1995;18:1417–1426.

[70] Iijima Y, Nagai T, Mizukami M, et al. Beating is necessary for transdifferentiation of skeletal muscle-derived cells into cardiomyocytes. *Faseb J.* 2003;17:1361–1363.

[71] Ghostine S, Carrion C, Souza LC, et al. Long-term efficacy of myoblast transplantation on regional structure and function after myocardial infarction. *Circulation.* 2002;106:131–136.

[72] Hagege AA, Carrion C, Menasche P, et al. Viability and differentiation of autologous skeletal myoblast grafts in ischaemic cardiomyopathy. *Lancet.* 2003;361:491–492.

[73] Menasche P, Hagege AA, Vilquin JT, et al. Autologous skeletal myoblast transplantation for severe postinfarction left ventricular dysfunction. *J Am Coll Cardiol.* 2003;41:1078–1083.

[74] Atkins BZ, Lewis CW, Kraus WE, et al. Intracardiac transplantation of skeletal myoblasts yields two populations of striated cells in situ. *Ann Thorac Surg.* 1999;67:124–129.

[75] Taylor DA, Atkins BZ, Hungspreugs P, et al. Regenerating functional myocardium: improved performance after skeletal myoblast transplantation. *Nat Med.* 1998;4:929–933.

[76] Rangappa S, Fen C, Lee EH, Bongso A, Wei ES. Transformation of adult mesenchymal stem cells isolated from the fatty tissue into cardiomyocytes. *Ann Thorac Surg.* 2003;75:775–779.

[77] Planat-Benard V, Menard C, Andre M, et al. Spontaneous cardiomyocyte differentiation from adipose tissue stroma cells. *Circ Res.* 2004;94:223–229.

[78] Gaustad KG, Boquest AC, Anderson BE, Gerdes AM, Collas P. Differentiation of human adipose tissue stem cells using extracts of rat cardiomyocytes. *Biochem Biophys Res Commun.* 2004;314:420–427.

[79] Toma C, Pittenger MF, Cahill KS, Byrne BJ, Kessler PD. Human mesenchymal stem cells differentiate to a cardiomyocyte phenotype in the adult murine heart. *Circulation.* 2002;105:93–98.

[80] Makino S, Fukuda K, Miyoshi S, et al. Cardiomyocytes can be generated from marrow stromal cells *in vitro. J Clin Invest.* 1999;103:697–705.

[81] Fukuda K. Development of regenerative cardiomyocytes from mesenchymal stem cells for cardiovascular tissue engineering. *Artif Organs.* 2001;25:187–193.

[82] Rangappa S, Entwistle JW, Wechsler AS, Kresh JY. Cardiomyocyte-mediated contact programs human mesenchymal stem cells to express cardiogenic phenotype. *J Thorac Cardiovasc Surg.* 2003;126:124–132.

[83] Barbash IM, Chouraqui P, Baron J, et al. Systemic delivery of bone marrow-derived mesenchymal stem cells to the infarcted myocardium. Feasibility, cell migration, and body distribution. *Circulation.* 2003;863–868.

[84] Shake JG, Gruber PJ, Baumgartner WA, et al. Mesenchymal stem cell implantation in a swine myocardial infarct model: engraftment and functional effects. *Ann Thorac Surg.* 2002;73:1919–1925.

[85] Tomita S, Mickle DA, Weisel RD, et al. Improved heart function with myogenesis and angiogenesis after autologous porcine bone marrow stromal cell transplantation. *J Thorac Cardiovasc Surg.* 2002;123:1132–1140.

[86] Yau TM, Tomita S, Weisel RD, et al. Beneficial effect of autologous cell transplantation on infarcted heart function: comparison between bone marrow stromal cells and heart cells. *Ann Thorac Surg.* 2003;75:169–176.

[87] Mangi AA, Noiseux N, Kong D, et al. Mesenchymal stem cells modified with Akt prevent remodeling and restore performance of infarcted hearts. *Nat Med.* 2003;9:1195–1201.

[88] Koc ON, Gerson SL, Cooper BW, et al. Rapid hematopoietic recovery after coinfusion of autologous-blood stem cells and culture-expanded marrow mesenchymal stem cells in advanced breast cancer patients receiving high-dose chemotherapy. *J Clin Oncol.* 2000;18:307–316.

[89] Nolta JA, Smogorzewska EM, Kohn DB. Analysis of optimal conditions for retroviral-mediated transduction of primitive human hematopoietic cells. *Blood.* 1995;86:101–110.

[90] Dunbar CE, Kohn DB, Schiffmann R, et al. Retroviral transfer of the glucocerebrosidase gene into CD34+ cells from patients with Gaucher disease: *in vivo* detection of transduced cells without myeloablation. *Hum Gene Ther.* 1998;9:2629–2640.

[91] Reese JS, Koc ON, Gerson SL. Human mesenchymal stem cells provide stromal support for efficient CD34+ transduction. *J Hematother Stem Cell Res.* 1999;8:515–523.

[92] Cousin B, Andre M, Arnaud E, Penicaud L, Casteilla L. Reconstitution of lethally irradiated mice by cells isolated from adipose tissue. *Biochem Biophys Res Commun.* 2003;301:1016–1022.

[93] Woodbury D, Schwarz EJ, Prockop DJ, Black IB. Adult rat and human bone marrow stromal cells differentiate into neurons. *J Neurosci Res.* 2000;61:364–370.

[94] Deng W, Obrocka M, Fischer I, Prockop DJ. *In vitro* differentiation of human marrow stromal cells into early progenitors of neural cells by conditions that increase intracellular cyclic AMP. *Biochem Biophys Res Commun.* 2001;282:148–152.

[95] Zhao LR, Duan WM, Reyes M, et al. Human bone marrow stem cells exhibit neural phenotypes and ameliorate neurological deficits after grafting into the ischemic brain of rats. *Exp Neurol.* 2002;174:11–20.

[96] Chen J, Zhang ZG, Li Y, et al. Intravenous administration of human bone marrow stromal cells induces angiogenesis in the ischemic boundary zone after stroke in rats. *Circ Res.* 2003;92:692–699.

[97] Li Y, Chen J, Chen XG, et al. Human marrow stromal cell therapy for stroke in rat: neurotrophins and functional recovery. *Neurology.* 2002;59:514–523.

[98] Azizi SA, Stokes D, Augelli BJ, DiGirolamo C, Prockop DJ. Engraftment and migration of human bone marrow stromal cells implanted in the brains of albino rats–similarities to astrocyte grafts. *Proc Natl Acad Sci USA.* 1998;95:3908–3913.

[99] Kopen GC, Prockop DJ, Phinney DG. Marrow stromal cells migrate throughout forebrain and cerebellum, and they differentiate into astrocytes after injection into neonatal mouse brains. *Proc Natl Acad Sci USA.* 1999;96:10711–10716.

[100] Prockop DJ, Azizi SA, Phinney DG, Kopen GC, Schwarz EJ. Potential use of marrow stromal cells as therapeutic vectors for diseases of the central nervous system. *Prog Brain Res.* 2000;128:293–297.

[101] Prockop DJ, Azizi SA, Colter D, et al. Potential use of stem cells from bone marrow to repair the extracellular matrix and the central nervous system. *Biochem Soc Trans.* 2000;28:341–345.

[102] Kang SK, Lee DH, Bae YC, et al. Improvement of neurological deficits by intracerebral transplantation of human adipose tissue-derived stromal cells after cerebral ischemia in rats. *Exp Neurol.* 2003;183:355–366.

[103] Kang SK, Jun ES, Bae YC, Jung JS. Interactions between human adipose stromal cells and mouse neural stem cells *in vitro*. *Brain Res Dev Brain Res.* 2003;145:141–149.

[104] Minasi MG, Riminucci M, De Angelis L, et al. The meso-angioblast: a multipotent, self-renewing cell that originates from the dorsal aorta and differentiates into most mesodermal tissues. *Development.* 2002;129:2773–2783.

[105] Sampaolesi M, Torrente Y, Innocenzi A, et al. Cell therapy of alpha-sarcoglycan null dystrophic mice through intra-arterial delivery of mesoangioblasts. *Science.* 2003;301:487–492.

[106] Miranville A, Heeschen C, Sengenes C, et al. Improvement of postnatal neovascularization by human adipose tissue-derived stem cells. *Circulation.* 2004;110:349–355.

[107] Rehman J, Traktuev D, Li J, et al. Secretion of angiogenic and antiapoptotic factors by human adipose stromal cells. *Circulation.* 2004.

[108] Horwitz EM, Prockop DJ, Gordon PL, et al. Clinical responses to bone marrow transplantation in children with severe osteogenesis imperfecta. *Blood.* 2001;97:1227–1231.

[109] Horwitz EM, Prockop DJ, Fitzpatrick LA, et al. Transplantability and therapeutic effects of bone marrow-derived mesenchymal cells in children with osteogenesis imperfecta. *Nat Med.* 1999;5:309–313.

[110] Morizono K, De Ugarte DA, Zhu M, et al. Multilineage cells from adipose tissue as gene delivery vehicles. *Hum Gene Ther.* 2003;14:59–66.

CHAPTER 7

UNCOMMITTED PROGENITORS IN CULTURES OF BONE MARROW-DERIVED MESENCHYMAL STEM CELLS

J. J. MINGUELL, A. ERICES AND W. D. SIERRALTA

Programa de Terapias Celulares, INTA, Universidad de Chile, Santiago, Chile

1. INTRODUCTION

Despite abundant *in vitro* data related with culture conditions and differentiation potential of *ex vivo* expanded mesenchymal stem cells (MSC) [1–5], there are still few comprehensive data on whether all expanded cells in a culture share a unique proliferative and differentiation potential. The accumulated knowledge, mainly after clonal studies of MSC [6–7], has indicated that cultures of MSC contain several categories of progenitors exhibiting divergent proliferative and differentiation properties. Thus, a proliferative hierarchy has been postulated to occur between a putative mesenchymal stem cell and its terminally differentiated progeny, including a heterogeneous population of intermediate committed progenitors [8–11]. The above is in consonance with the concept that as a stem cell proceeds towards a mature phenotype (s), its stemness gradually changes (self-renewal decreases and commitment increases) [12–13].

In vivo, and under steady-state conditions it has been demonstrated the existence of a quiescent mesenchymal stem cells [14], which upon entering into cell cycle commit and terminal differentiate into the various mesenchymal lineages [8–15]. Recent data have described the isolation of a homogeneous population of mesenchymal progenitors from adult human bone marrow, which may correspond to the *in vivo* uncommitted MSC. Cells thus isolated are non-cycling, constitutively express telomerase activity *in vivo*, exhibit an extensive proliferation potential after exposure to serum and a capacity for differentiation into bone, cartilage and adipose tissue [16]. The constitutive expression of telomerase activity *in vivo* by uncommitted MSC seems to represents a distinctive trait for these progenitors, since the transduction of *ex vivo* cultured human MSC with telomerase reverse transcriptase create immortal cell lines. These telomerase positive cells did not form foci in soft agar, have a normal karyotype and differentiate into osteoblasts and chondrocytes [17].

J.A. Nolta (ed.), Genetic Engineering of Mesenchymal Stem Cells, 127–133.

Altogether, these and other studies have strengthened the notion that bone marrow is the site of residency of an uncommitted mesenchymal stem cell, which upon expression of its self-renewal and multidifferentiation potential, commands the continual replenishing of a given supply of mesenchymal cells during the entire lifespan of an organism, both at steady-state and altered conditions [18].

2. THE SEARCH FOR UNCOMMITTED PROGENITORS IN EXPANDED MSC CULTURES

A relevant issue in understanding the biology of MSC and consequently in their use in novel cellular therapies, relates with the concept on whether in a "culture dish of expanded MSC," exists of a subset of uncommitted and self renewing progenitors. In most reports the presence of such progenitors (which may represent the *ex vivo* counterpart of *in vivo* MSC) has been considered an obvious and indisputable matter.

Expanded cultures of human bone marrow-derived MSC have proven to be predominantly homogeneous, as based on morphology, immunophenotype and response to differentiation stimuli [2–5]. When the cell cycle status of these cultures was analysed, it was found that approximately 60–80% of cells were standing at the Go/G1 phase, of which between 5% and 20% were quiescent Go cells [4]. Therefore, the strategy used by these authors to isolate the Go subset, was to treat expanded cultures of MSC with a proper antimetabolite exhibiting a maximum killing activity for cells in cycle. Initial trials showed that the pyrimidine analogue 5-fluorouracil (5-FU) [18], proved to be the most effective drug in selecting a "healthy" population of quiescent, Go cells [19]. Based on the observation that the rate of expansion and yield of multipotential progenitors are inversely related with plating density and incubation time [20], a minor population of quiescent progenitors was also isolated from cultures of bone marrow-derived MSC cells [21–22]. Under these conditions, two populations of cells were separated from low density-seeded cultures of MSC. The minor population was formed by small and agranular quiescent cells with a low capacity to generate colonies (RS-1 cells), while the most abundant population contained fast growing committed cells (mMSC's). By studying a precursor-product relationship between RS-1 and mMSC's cells, it was demonstrated that mMSC's arose from quiescent uncommitted RS-1 cells after their entry in cycle [21]. Moreover, the existence of quiescent progenitors has been also revealed by the use of clonal cultures of bone marrow-derived MSC [6, 7]. Accordingly, and despite the cellular stress associated with *ex vivo* manipulation procedures, cultures of expanded MSC still contain a limited subset of cells exhibiting properties often assigned to stem cells or early progenitors, like quiescence and a lack of commitment.

3. CHARACTERISTICS OF UNCOMMITTED PROGENITORS IDENTIFIED IN EXPANDED CULTURES OF MESENCHYMAL STEM CELLS

As examined by phase contrast microscopy, progenitors isolated after the use of 5-FU, appeared as small stellate cells with an overt flattening. In turn, transmission electron microscopic analysis revealed as main characteristic of these cells, the presence of a thin layer of cytoplasm surrounding the nucleus, a low number of free ribosomes and lysosomes, few RER, several small dark intermediate-type mitochondrion and predominance of heterochromatin adjacent to the inner nuclear membrane [23, 24]. In turn, cell cycle analysis demonstrated that cultures of uncommitted progenitors consist of a large fraction of cells (94%) at the Go/G1 phase, being the rest distributed along the S+ G2/M phases. Since low or negligible RNA content as well as low mitochondrial activity are distinctive traits for quiescent cells [25–27], the subset of Go cells in the G_0/G1 fraction was assessed after staining sequentially with acridine orange and rhodamine 123. Results showed that 87% and 70% of cells displayed a RNA^{low}/DNA^{2n} and a Pi^-/Rho^{low} phenotype, respectively [19, 24]. Further evidence for the Go condition of these cells was obtained by analysis of gene expression of the anti-proliferative marker, ornithine decarboxylase (ODC) antizyme [19, 28]. Thus, these studies have shown that cultures of expanded MSC contain a minor population of cells displaying morphological and proliferative features dissimilar to those of the vast majority of cells present in these cultures [3, 4]. Similarly, after plating early passage MSC at low densities, Colter et al. observed the emergence of small and agranular quiescent cells (RS-1 cells) [21], which may correspond to the quiescent cells isolated after the 5-FU procedure [22, 23].

Properties of 5-FU-isolated quiescent progenitors, like no expression of commitment markers and unresponsiveness to differentiation stimuli, have been disclosed shortly after isolation and in the absence of fetal bovine serum (FBS). After incubation in culture medium containing FBS, cells express $PPAR\chi 2$ and Cbfa-1, recognize the effect of differentiation stimuli and express a self-renewal capacity [19, 23]. These effects of FBS on quiescent, uncommitted MSC are not without precedent. The differentiation of rat mesenchymal cells into a neural lineage occurs only in the absence of serum, since nestin expression (a requisite for differentiation) is inhibited by cell exposure to FBS [29]. In the same way recent studies have shown that uncommitted mesenchymal progenitors isolated from unprocessed bone marrow, once exposed to FBS switch from a $CD45^{med,low}$/$CD34^{low}$ into a $CD45^-$/$CD34^-$ phenotype [30].

Therefore, it seems that uncommitted progenitors during their (experimental) displacement from the bone marrow microenvironment to the tissue culture incubator, interpret exposure to fetal serum not as a widespread mitogenic stimulus but as an injury signal (for *ex vivo* mesengenesis?), as it occurs with skin fibroblasts [31]. At a molecular basis, the effect of serum may be related to changes in the secretion levels of dickkopf-1 (Dkk-1) [32, 33], an inhibitor of the Wnt signalling [34]. Thus, it seems that the observed effects of serum in quiescent MSC may be related to a modulation in the transition G_0- cell cycle, via the Dkk-1/Wnt/beta-catenin signalling pathway, as it occurs in other systems [35].

Flow cytometric studies have shown that uncommitted progenitors isolated from expanded MSC cultures express at least, α-ASMA, CD105, CD51/CD61(αvβ3), CD51/b5 (αvβ5), CD117, the adenovirus attachment receptor (CAR), cytokeratin 18 [19] as well as CD31, CD38 and CD90 [21]. An interesting and controversial issue associated with the immunophenotype of either *in vivo* or *ex vivo* uncommitted MSC is related with the expression of CD45, a typical marker of the hematopoietic lineage. *Ex vivo* isolated uncommitted progenitors express a stable $CD45^-$ phenotype through subsequent passages [19, 21], however *in vivo* uncommitted progenitors (from unprocessed bone marrow) are characterised by expressing a CD45^{med,low} immunophenotype [30].

4. EFFECT OF GROWTH FACTORS ON THE DEVELOPMENT OF UNCOMMITTED MESENCHYMAL PROGENITORS

While several studies have addressed the ability of growth factors and/or cytokines to initiate and support the clonogenic growth of *ex vivo* expanded marrow-derived mesenchymal progenitors [36–38], scarce information is available on the involvement of specific growth factors or cytokines in the development of uncommitted MSC. Indirect data have suggested that bFGF is able to sustain self-renewal of MSC and maintain cells in a more immature state [39, 40]. To gain further insight into the developmental effect of growth factors in *ex vivo*-isolated uncommitted progenitors, we assessed in these cells the expression of a selected number of growth factor receptors as well as the mitogenic effect evoked by the respective factor. Uncommitted MSC express c-kit (CD117), PDGFRa, gp130 and FGFR1, but do not express IL-6R, GM-CSFR (a and b chains), LIFR and VEGFR. As predicted, when cells were incubated with LIF, GM-CSF or IL-6 no proliferative effects were observed [19, 23, 41], however after exposure to SCF or PDGFa, an increase in cell number was observed. In the later case, the proliferative effect was associated with the retention of the quiescent condition of the cells [19]. The competence of PDGFa to support the expansion and maintenance of uncommitted MSC is not unexpected since the growth factor also expands the pool of other immature cells, like neuron and glial progenitors [42, 43]. When the mitogenic effect of bFGF was examined it was observed that the growth factor did not evoke a proliferative response in uncommitted cells, rather it produced a moderate effect in sustaining their quiescent condition [24], an effect already claimed to be specific for bFGF [39, 40].

5. FUTURE DIRECTIONS IN THE BIOLOGY OF UNCOMMITTED MESENCHYMAL STEM CELLS

The last years have been the scene of a substantial improvement in our understanding of the biology and the potential clinical utilisation of adult MSC. Many aspects related to the presence and properties of uncommitted mesenchymal progenitors both in expanded cultures of MSC as well as in unprocessed bone marrow, are now better established. However, information dealing with other aspects of the biology of early mesenchymal progenitors still remain obscure, such as: (i) the mechanisms involved in MSC migration and mobilization. Evidence indicates that marrow-resident mesenchymal progenitors are mobilized and detected in fetal or adult blood [45, 46]. Similarly, when the demand for mesenchymal precursors in an injured tissue exceeds that of resident precursors, marrow progenitors are mobilized and participate in the regeneration of the damaged tissue [47]. It is not known whether cytokines, chemokine receptor/ligand pairs or other factors are involved in MSC mobilization, however SDF-1, the ligand for CXCR4 is produced by bone marrow stromal cells [48], (ii) a microenvironment (s) for commitment and maturation of uncommitted MSC has not been delineated.

Mesenchymal progenitors in adult tissues are potentially capable of differentiating into several lineages, however under steady state conditions they are inhibited to express this potential by microenvironmental factors. Moreover, when the microenvironment changes considerably (i.e. pathology), the inhibiting condition is removed and cells can differentiate and mature into a specific lineage [49, 50]. In addition, it is known that mesenchymal progenitors delivered via systemic infusion engraft into various tissues with a low efficiency [1–5] and therefore may not produce relevant, robust and durable clinical effects. It has not been established whether "manipulation of a particular microenvironment" will increase the engraftment of transplanted stem/progenitor cells. Thus, a better understanding of mobilization as well as of the nature of the cellular and molecular microenvironment (s) [52] will, hopefully, extend the field of therapeutic applications of mesenchymal stem cells.

REFERENCES

[1] Pereira RF, Halford KW, O'Hara MD, et al. Cultured adherent cells from marrow can serve as long-lasting precursor cells for bone, cartilage, and lung in irradiated mice. *Proc Natl Acad of Sci USA.* 1995;92:4857–4861.

[2] Majumdar MK, Thiede MA, Mosca JD, Moorman M, Gerson SL. Phenotypic and functional comparison of cultures of marrow-derived mesenchymal stem cells (MSCs) and stromal cells. *J Cell Physiol.* 1998;176:57–66.

[3] Pittenger MF, Mackay AM, Beck SC, et al. Multilineage potential of adult human mesenchymal stem cells. *Science.* 1999;284:143–147.

[4] Conget PA, Minguell JJ. Phenotypical and functional properties of human bone marrow mesenchymal progenitor cells. *J Cell Physiol.* 1999;81:67–73.

[5] DiGirolamo CM, Stokes D, Colter D, Phinney DG, Class R, Prockop DJ. Propagation and senescence of human marrow stromal cells in culture: a simple colony-forming assay identifies samples with the greatest potential to propagate and differentiate. *Br J Haematol.* 1999;107:275–281.

[6] Dennis JE, Merriam A, Awadallah A, Yoo JU, Johnstone B, Caplan AI. A quadripotential mesenchymal progenitor cell isolated from the marrow of an adult mouse. *J Bone Miner Res.* 1999;14:700–709.

[7] Muraglia A, Cancedda R, Quarto R. Clonal mesenchymal progenitors from human bone marrow differentiate *in vitro* according to a hierarchical model. *J Cell Sci.* 2000;113:1161–1166.

[8] Caplan AI. The mesengenic process. *Clin Plast Surg.* 1994;21:429–435.

[9] Minguell JJ, Erices A, Conget P. Mesenchymal stem cells. *Exp Biol Med.* 2001;226:507–520.

[10] Bordignon C, Carlo-Stella C, Colombo MP, et al. Cell therapy: achievements and perspectives. *Haematologica.* 1999;84:1110–1149.

[11] Bruder SP, Horowitz MC, Mosca JD, Haynesworth SE. Monoclonal antibodies reactive with human osteogenic cell surface antigens. *Bone.* 1997;21:225–235.

[12] Morrison SJ, Shah NM, Anderson DJ. Regulatory mechanism in stem cell biology. *Cell.* 1997;88:287–298.

[13] Asahara TC, Kalka C, Isner JM. Stem cell therapy and gene transfer for regeneration. *Gene Ther.* 2000;7:451–457.

[14] Castro-Malaspina HC, RE Gay, Resnick G, et al. Characterization of human bone marrow fibroblast colony-forming cells (CFU-F) and their progeny. *Blood.* 1980;56:289–301.

[15] Friedenstein AJ, Chailakhyan RK, Latzinik NV, Panasyuk A, Keilis-Borok IV. Stromal cells responsible for transferring the microenvironment of hemopoietic tissues: cloning *in vitro* and retransplantation *in vivo*. *Transplantation.* 1974;17:331–340.

[16] Gronthos S, Zannettino AC, Hay SJ, et al. Molecular and cellular characterisation of highly purified stromal stem cells derived from human bone marrow. *J Cell Sci.* 2003;116:1827–1835

[17] Mihara K, Imai C, Coustan-Smith E, et al. Development and functional characterization of human bone marrow mesenchymal cells immortalized by enforced expression of telomerase. *Br J Haematol.* 2003;120:846–849.

[18] Grem JL, Hoth DF, Hamilton JM, King SA, Leyland-Jones B. Overview of current status and future direction of clinical trials with 5-fluorouracil in combination with folinic acid. *Cancer Treat Rep.* 1987;71:1249–1264.

[19] Conget PA, Allers C, Minguell JJ. Identification of a discrete population of human bone-marrow derived mesenchymal cells exhibiting properties of uncommitted progenitors. *J Hematother Stem Cell Res.* 2001;10:749–758.

[20] Mets T, Verdonk G. Variations in the stromal cell population of human bone marrow during aging. *Mech Ageing Dev.* 1981;15:41–49.

[21] Colter DC, Class R, DiGirolamo CM, Prockop DJ. Rapid expansion of recycling stem cells in cultures of plastic-adherent cells from human bone marrow. *Proc Natl Acad Sci USA.* 2000;97:3213–3218.

[22] Pochampally RR, Smith JS, Sekiya I, Colter DC, Prockop DJ. Selection of rapidly self renewing cells from cultures of adult stem cells from bone marrow stroma. In: Grisolia S, Minana MD, Bendala-Tufanisco E, eds. *Mesenchymal Stem Cells: Biology and potential clinical uses.* . Valencia, Spain: Fundacion Valenciana de Estudios Avanzados; 2003:70–84.

[23] Minguell JJ, Erices A, Conget PA, et al. Mesenchymal (uncommitted) Stem Cells. In: Grisolia S, Minana MD, Bendala-Tufanisco E, eds. *Mesenchymal Stem Cells: Biology and potential clinical uses.* Valencia, Spain: Fundacion Valenciana de Estudios Avanzados; 2003:11–36.

[24] Benavente C, Sierralta WD, Conget PA, Minguell JJ. Subcellular distribution and mitogenic effect of basic fibroblast growth factor in mesenchymal uncommitted stem cells. *Growth Factors.* 2003;21(2):87–94.

[25] Kim M, Cooper DD, Hayes SF, Spangrude GJ. Rhodamine-123 staining in hematopoietic stem cells of young mice indicates mitochondrial activation rather than dye efflux. *Blood.* 1998;91:4106–4111.

[26] Ratajczak MZ, Pletcher CH, Marlicz W, et al. CD34$^+$, kit$^+$, rhodamine123^{low} phenotype identifies a marrow cell population highly enriched for human hematopoietic stem cells. *Leukemia.* 1998;12:942.

[27] Juan G, Darzynkiewicz Z. Cell cycle analysis by flow and laser scanning cytometry. In: JE Celis, ed. *Cell Biology: A Laboratory Handbook.* Vol 1. San Diego, CA: Academic Press; 1998:261–272.

[28] Iwata S, Sato Y, Asada M, et al. Anti-tumor activity of antizyme which targets the ornithine decarboxylase (ODC) required for cell growth and transformation. *Oncogene.* 1999;18:165–172.

[29] Wislet-Gendebien S, Leprince P, Moonen G, Rogister B. Regulation of neural markers nestin and GFAP expression by cultivated bone marrow stromal cells. *J Cell Sci.* 2003;116:3295–3302.

[30] Deschaseaux F, Gindraux F, Saadi R, Obert L, Chalmers D, Herve P. Direct selection of human bone marrow mesenchymal stem cells using an anti-CD49a antibody reveals their CD45med,low phenotype. *Br J Haematol.* 2003;122:506–517.

[31] Iyer VR, Eisen MB, Ross DT, et al. The transcriptional program in the response of human fibroblasts to serum. *Science.* 1999;283:83–87.

[32] Gregory CA, Singh H, Perry AS, Prockop DJ. The Wnt signaling inhibitor dickkopf-1 is required for reentry into the cell cycle of human adult stem cells from bone marrow. *J Biol Chem.* 2003;278:28067–28078.

[33] Prockop DJ, Gregory CA, Spees JL. One strategy for cell and gene therapy: harnessing the power of adult stem cells to repair tissues. *Proc Natl Acad Sci USA.* 2003;100:11917–23.

[34] Shou J, Ali-Osman F, Multani AS, Pathak S, Fedi P, Srivenugopal KS. Human Dkk-1, a gene encoding a Wnt antagonist, responds to DNA damage and its overexpression sensitizes brain tumor cells to apoptosis following alkylation damage of DNA. *Oncogene.* 2002;21:878–889.

[35] Boyden LM, Mao J, Belsky J, et al. High bone density due to a mutation in LDL-receptor-related protein 5. *N Engl J Med.* May 16, 2002;346(20):1513–1521.

[36] Gronthos S, Simmons PJ. The growth factor requirements of STRO-1-positive human bone marrow stromal precursors under serum-deprived conditions *in vitro. Blood.* 1995;85:929–939.

[37] Kuznetzov SA, Friedenstein AJ, Robey PG. Factors required for bone marrow fibroblast colony formation *in vitro. Br J Haematol.* 1997;97:561–569.

[38] Oliver LJ, Rifkin DB, Gabrilove J, Hannocks MJ, Wilson EL. Long-term culture of bone marrow stromal cells in the presence of basic fibroblast growth factor. *Growth Factors.* 1990;3:231–239

[39] Martin I, Muraglia A, Campanile G, Cancedda R, Quarto R. Fibroblast growth factor-2 supports *ex vivo* expansion and maintenance of osteogenic precursors from human bone marrow. *Endocrinology.* 1997;138:4456–4461.

[40] Tsutsumi S, Shimazu A, Miyazaki K, Pan H, Koike C, Yoshida E, Takagishi K, Kato Y. Retention of multilineage differentiation potential of mesenchymal cells during proliferation in response to FGF. *Biochem Biophys Res Commun.* 2001;288:413–422.

[41] Erices A, Conget P, Rojas C, Minguell JJ. Gp130 activation by soluble interleukin-6 receptor/interleukin-6 enhances osteoblastic differentiation of human bone marrow-derived mesenchymal stem cells. *Exp Cell Res*. 2002;280:24–32.

[42] Erladsson A, Enarsson M, Forsberg-Nilsson K. Immature neurons from CNS stem cells proliferate in response to platelet-derived growth factor. *J Neurosci*. 2001;21:3483–3491.

[43] Smith J, Ladi E, Mayer-Proschel M, Noble M. Redox state is a central modulator of the balance between self-renewal and differentiation in a dividing glial precursor cell. *Proc Natl Acad Sci*. 2000;97:10032–10037.

[44] Fernández M, Simon V, Herrera G, Cao C, Del Favero H, Minguell JJ. Detection of stromal cells in peripheral blood progenitor cell collections from breast cancer patients. *Bone Marrow Transplant*. 1997;20:265–271.

[45] Zvaifler NJ, Marinova-Mutafchieva L, Adams G, et al. Mesenchymal precursor cells in the blood of normal individuals. *Arthritis Res*. 2000;2:477–488.

[46] Erices A, Conget P, Minguell JJ. Mesenchymal progenitor cells in human umbilical cord blood. *Br J Haematol*. 2000;109:235–242.

[47] Ferrari G, Cusella-De Angelis G, Coletta M, et al. Muscle regeneration by bone marrow-derived myogenic progenitors. *Science*. 1998;279:1528–1530.

[48] Fruehauf S, Seggewiss R. It's moving day: factors affecting peripheral blood stem mobilization and strategies for improvement. *Br J Haematol*. 2003;122:360–375.

[49] Cancedda R, Bianchi G, Derubeis A, Quarto R. Cell therapy for bone disease: a review of current status. *Stem Cells*. 2003;21:610–619.

[50] Sordella R, Jiang W, Chen GC, Curto M, Settleman J. Modulation of Rho GTPase signaling regulates a switch between. *Cell*. 2003;113:147–158.

[51] Erices AA, Allers CI, Conget PA, Rojas CV, Minguell JJ. Human cord blood-derived mesenchymal stem cells home and survive in the marrow of immunodeficient mice after systemic infusion. *Cell Transplant*. 2003;12(in press)

[52] Watt FM, Hogan BL. Out of Eden: stem cells and their niches. *Science*. 2000;287:1427–1430.

CHAPTER 8

BONE MARROW MESENCHYMAL STEM CELL TRANSPLANTATION FOR CHILDREN WITH SEVERE OSTEOGENESIS IMPERFECTA

E. M. HORWITZ[1,2] AND P. L. GORDON[1]

Divisions of Stem Cell Transplantation[1] and Experimental Hematology[2]
Department of Hematology-Oncology, St Jude Children's Research Hospital,
Memphis, TN

1. INTRODUCTION

In 1957, Dr. E. Donnell Thomas and colleagues transplanted freshly harvested unmanipulated bone marrow into six adult patients with leukemia, prepared with radiation and chemotherapy to destroy the leukemic cells, in an effort to transplant the putative hematopoietic stem cells. [1] that, in animal models, had been shown to regenerate hematopoiesis after lethal irradiation [2–4]. Although a cell phenotype for these hematopoietic repopulating cells would not be described for another 27 years [5], the hypothesis in 1957 was that these unidentified, rare cells within donor bone marrow would home to the patient's marrow space and reconstitute hematopoiesis, rescuing the patient from the lethal effects of the radiation intended to eradicate the leukemia. Only a single patient showed any engraftment of donor cells, and that was transient; however, important biomedical principles were proven, and this bold clinical trial ushered in the era of bone marrow transplantation. In 1990, Dr. Thomas was awarded the Nobel Prize in Medicine for his pioneering work.

The field of bone marrow mesenchymal stem cells (MSCs), the subject of this entire volume, began with Alexander Friendenstein's critical observation that bone marrow contains adherent, fibroblastic cells which can differentiate to cells with an osteogenic phenotype [6]. He continued to characterize these cells and found that they could proliferate in culture and differentiate, not only to bone, but also to chondrocytes and adipocytes [7–11]. Thus, these marrow stromal cells, which can support hematopoiesis, can self-renew and differentiate to at least three tissues, meeting the definition of a "stem cell" proposed earlier by Till and McCulloch [12]. Maureen Owen proposed the concept of a stromal stem cell [13, 14], but the work of Arnold Caplan and colleagues is most

J.A. Nolta (ed.), Genetic Engineering of Mesenchymal Stem Cells, 135–150.

responsible for advancing the study of marrow stromal cells as cellular therapy in the early 1990's. Caplan popularized the term, "mesenchymal stem cell," to identify these multipotent adherent fibroblastic progenitors in bone marrow [15].

The idea of using MSCs as cellular therapy for systemic disorders was brought to the forefront by the seminal report of Prockop and colleagues in 1995, who tracked the fate of genetically marked MSCs after intravenous infusion in a mouse model [16]. These investigators found that the infused MSCs engrafted in bone, cartilage and lung parenchyma in both alveoli and bronchi, suggesting the potential of MSC therapy for genetic disorders affecting these tissues.

2. RATIONALE FOR BONE MARROW TRANSPLANTATION THERAPY FOR CHILDREN WITH SEVERE OSTEOGENESIS IMPERFECTA

Bone marrow transplantation (BMT) is accepted as an effective therapeutic modality for genetic and acquired diseases of the hematopoietic system. This cellular therapy is based on the knowledge that bone marrow contains hematopoietic stem cells, which can engraft and differentiate to blood cells. With our recognition that bone marrow also contains mesenchymal stem cells that can differentiate to several mature mesenchymal cell types, bone marrow transplantation, as a means of transplanting both hematopoietic and mesenchymal cells, could, in principle, be used to treat disorders of mesenchymal tissues as well as those of the blood.

Osteogenesis Imperfecta (OI) is a genetic disorder of mesenchymal cells in which generalized osteopenia leads to bony deformities, excessive fragility with fractures, and markedly short statue. The underlying defect is a mutation in one of the two genes encoding type I collagen, the primary structural protein of bone [17]. There is an enormous variability in the severity of clinical phenotype among patients with OI, and physicians have classified the patients into four "types" [18]. Type I is the most mild phenotype and these patients are not easily recognized as having the disorder. Type II is the most severe phenotype with nearly all children dying of their disease or related complications prior to their first birthday. Type IV patients are considered moderate, while type III patients, the so-called "progressive deforming OI," are the most severely affected group that routinely survive infancy, while suffering severe deformities and painful fractures. These type III patients typically will attain a final stature of about $3-3^{1}/_{2}$ feet. Historically, the life expectancy of the type III patients was quite short. However, with advances in supportive care, they usually now live well into adulthood.

In the middle of the 1990's, much work had been published on MSCs [15], and autologous MSCs had been safely infused into adult patients undergoing autologous bone marrow transplantation for breast cancer [19]; however, the field of MSC biology was really in its infancy. While few doubted the existence of a marrow mesenchymal stem cell, many controversies surrounded our understanding of the nature of this putative stem cell [20]. For example, the SH2+ SH3+ adherent marrow cells were well established as a candidate mesenchymal stem cell [21] with osteogenic potential. However, Long and colleagues had described CD34- nonadherent light density marrow cells with robust osteogenic differentiation capacity in vitro [22, 23]. Over the past few years, other marrow cells with mesenchymal stem cell characteristics have been reported.

RS cells [24, 25] and MAPCs [26] are adherent marrow cells and SP cells [27] are nonadherent cells, all of which have been shown to differentiate to osteogenic cells. Thus, in retrospect, our initial strategy to transplant whole bone marrow as a means of transplanting the marrow mesenchymal cell with osteogenic capacity seems quite prudent, and is further justified by more recent data.

The most fundamental question is why we chose osteogenesis imperfecta as a model disorder to validate the principle of marrow mesenchymal stem cell transplantation. First, the osteogenic differentiation capacity of marrow mesenchymal cells was well established, suggesting a mesenchymal disorder of bone as an ideal disorder. Second, studies of the parents of probands with lethal OI indicated that some parents were mosaic for the same mutation that produced severe OI in the offspring [28]. The mosaic parents were asymptomatic, even though the ratio of mutated to normal alleles in some tissues approached the value of 1:1 seen in the tissue of their affected offspring. This finding indicates that the severity of the disease is dependent, in part, on the relative balance between the rate of synthesis of mutated and normal collagen. Indeed, different lines of transgenic mice that expressed various levels of the same mutated COL1A1 gene (one of the two genes that code for collagen type I) showed a range of OI manifestations [29]. Therefore, even low levels of mesenchymal stem/progenitor cell engraftment may be sufficient to produce a shift in the balance between the synthesis of mutated and normal collagen, thereby benefiting the children by converting a severe OI phenotype to a less severe one. Finally, preclinical data indicated that intravenously infused MSCs can migrate to and become incorporate into bone in animal models [16, 30]. In fact, in a model of OI, mesenchymal engraftment produced appreciable improvement in the disease phenotype [30].

The goal of our research is to develop widely applicable mesenchymal cell therapy that will reduce the severity of a child's phenotype. For such novel research, only children with a severe phenotype would be eligible. In our first clinical trial we transplanted children with unmanipulated bone marrow and then evaluated the patients for donor mesenchymal engraftment, improvement in bone histology, and for indications of clinical benefit [31]. In the second protocol, the same cohort of children were transplanted with isolated MSCs derived from the bone marrow of the original donors, in an effort to "boost" the mesenchymal activity in the patients' bone [32]. In this study, we used a double gene marking strategy, pioneered by Brenner and colleagues [33], to better understand the relevant biology of human MSC transplantation, which would then lead to improvements in mesenchymal cell therapy for children with OI.

3. TREATMENT OF CHILDREN WITH SEVERE OSTEOGENESIS IMPERFECTA BY BONE MARROW TRANSPLANTATION: A PILOT STUDY

3.1. Transplantation

In the first clinical trial, five children with severe deforming OI (type III) (Figure 8.1) were intravenously infused with unmanipulated bone marrow from HLA-identical or single antigen mismatched siblings, after they had received ablative conditioning therapy. The diagnosis of OI was based on the identification of a genetic mutation

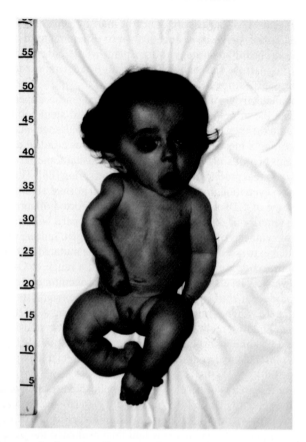

Figure 8.1. Photograph of a patient illustrating the typical features of severe deforming (type III) osteogenesis imperfecta. This 13-month-old girl has relative macrocephaly, triangular facies and blue sclera. Although not evident here, she also has malformed teeth indicative of dentinogenesis imperfecta. Her extremities are shorter than normal, with mild curvature of the arms and marked angulation deformities of the lower legs. On radiographs (not shown), the bones appeared thin and osteopenic with prominent curvatures not seen by physical examination. The humerus had an angulation of about 30° and the lower legs one of about 100°. There was also evidence of old fractures. The thoracic cage was small, and the ribs were malformed.

associated with type III OI, and the presence of physical features indicative of poor bone growth and development. The conditioning regimens consisted of busulfan and cyclophosphamide in three patients; busulfan, cyclophosphamide, and cytarabine in one; and busulfan, cyclophosphamide and moderate dose (900 cGy) fractionated total body irradiation in one patient who had an HLA DRβ1 allele mismatch with his sibling. Chemoprophylaxis against graft versus host disease consisted of cyclosporine in all children.

The first, and quite important, issue was to determine the toxicity of BMT for children with OI. Although mesenchymal stem cell transplantation *per se* was unlikely to result in unusual adverse events, there is little data to predict adverse outcomes related to the conditioning regimen in this patient population. In our cohort of children, there was no unusual or unacceptable toxicity. One child developed sepsis and transient pulmonary in-sufficiency and another developed acute graft versus host disease. All toxicities resolved and all children were discharged from the hospital in good medical condition.

3.2. Engraftment

All five patients showed engraftment of hematopoietic cells. One patient had a mixed hematopoietic chimerism (21% donor blood cells) that was stable for more than 3 years. In the other four patients, >99% of blood cells were of donor origin. Approximately 3 months after BMT, we obtained a bone biopsy from the iliac wing and harvested osteoblasts from bone explants in the laboratory [34]. After culture expansion, the adherent cells had typical osteoblast morphology, expressed alkaline phosphatase, and produced stainable extracellular matrix. We then verified the absence of hematopoietic contamination by flow cytometry and analyzed the confirmed osteoblasts for the presence of donor cells.

If the donor-recipient pair was sex mismatched, we assessed engraftment using fluorescence in situ hybridization (FISH) for the X and Y chromosomes. If the donor-recipient pair was of the same sex, we used an analysis for the variable number of tandem repeat (VNTR) DNA segments. We were able to demonstrate donor osteoblast engraftment in three of five patients, ranging from 1.2 to 2.0% of the total number of osteoblasts isolated from the iliac wing biopsy (Figure 8.2). Due to the small size of the patients, the bone specimens were, of necessity, quite small, compelling us to culture expand the bone cells prior to analysis. However, only mitotically active osteoblasts will proliferate in tissue culture, so that our engraftment analysis evaluates the fraction of donor osteoblasts in culture. In typical bone specimens, there are 10-fold more osteocytes than osteoblasts [35]. Thus, the actual mesenchymal engraftment may exceed the chimerism determined from cultured osteoblasts, and data in our laboratory from murine transplant studies supports the notion that the total osteogenic engraftment is greater than that measured by analysis of cultured osteoblasts (unpublished observation). Finally, bone is quite heterogeneous and it is conceivable that the epiphysis of long bones would contain a greater fraction of donor cells than the iliac wing, which is the standard and most surgically accessible site, for biopsy in the evaluation of metabolic bone disease.

3.3. Bone histology

If mesenchymal engraftment were to decrease the severity of the patient's phenotype, we hypothesized that engraftment should be associated with an improvement in the microscopic structure of the bone. Specimens of trabecular bone were obtained before BMT and again, from a distinct site, 6 months after BMT (Figure 8.3). Before transplant, trabecular bone typically contained disorganized osteocytes, enlarged lacunae, and relatively few osteoblasts (Figure 8.3.A). The bone had the characteristic appearance of high bone turnover, including woven bone (Figure 8.3.E), which is characteristic of OI and other metabolic bone disorders. Fluorescence microscopy of the same

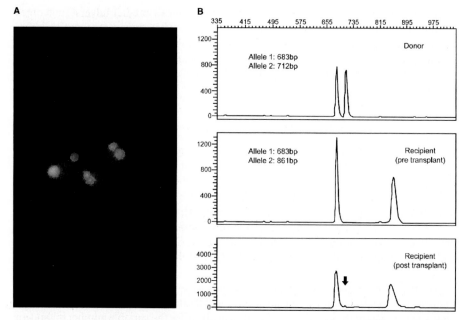

Figure 8.2. Analysis of osteoblast engraftment. (A) Fluorescence in situ hybridization (FISH) analysis of interphase nuclei from the cultured osteoblasts obtained from an iliac wing biopsy specimen 3 months after transplantation. Both X (middle and right), and Y (left) chromosomes are present in one of the cells from this female patient. Altogether, 1.5% of the cells studied were of donor (male) origin. Of the 500 female control cells counted, all demonstrated an XX pattern. (B) Electropherograms based on an analysis of DNA polymorphisms of the donor (top panel) and patient (middle panel) before transplantation, and of osteoblasts from the patient after transplantation (bottom panel). The peak indicated by the arrow represents about 2% donor cells.
(A color version of this figure is freely accessible via the website of the book: *http://www. springer.com/1-4020-3935-2*)

specimen showed distorted pattern of tetracycline labeling (Figure 8.3.C), consistent with the disorganized formation of new bone and poor mineralization. In contrast, similar specimens taken 6 months after transplant showed a reduced number of osteocytes, osteoblasts organized along the growing surface of bone, and evidence of lamellar bone formation (Figures 8.3.B and 8.3.F). Moreover, fluorescence microscopy showed a linear, single and double, tetracycline labeling pattern (Figure 8.3.D). Hence, we conclude that marrow mesenchymal engraftment in bone is associated with an improvement in the mechanism of bone formation and mineralization.

3.4. Clinical outcome

For the evaluation of the clinical outcome, we assessed three measures: growth velocity, bone mineral content, and fracture rate. These three parameters were chosen because they can be objectively evaluated, and because a therapy that can ameliorate these symptoms would certainly lessen the hardship faced by this population of patients.

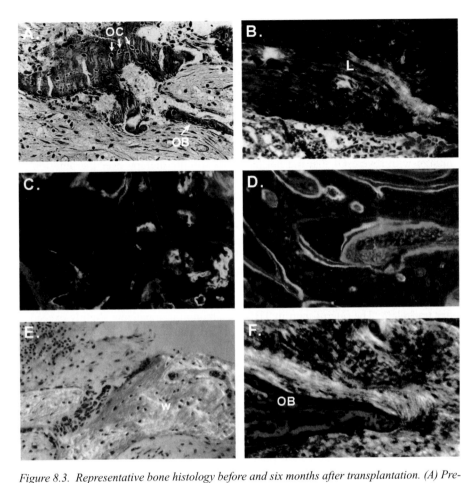

Figure 8.3. Representative bone histology before and six months after transplantation. (A) Pre-transplantation biopsy specimen of trabecular bone stained with Goldners-Masson trichrome. The calcified tissue appears blue-green, and the uncalcified tissue is red-brown. Numerous, randomly arranged osteocytes (OC) are sitting in large lacunae. Note also the peritrabecular marrow fibrosis, the paucity of osteoblasts relative to the post-transplantation specimens, and the incompletely calcified area of bone matrix. (B) Similarly prepared post-transplantation specimen, taken near the site shown in Figure 8.3.A. The number of osteocytes is reduced, and there is a small section of lamellar bone (L), suggesting normalization of the remodeling process. Magnification, 88X. (C) Fluorescence photomicrograph of the tetracycline-labeled trabecular bone specimen (same section as in Figure 8.3.A). The labeling is poorly defined, indicating disorganized formation of new bone and abnormal mineralization. (D) Contrasting post-transplantation specimen with definitive, crisp single and double tetracycline labeling, indicative of markedly improved new bone formation and mineralization. Magnification, 56X. (E) Pretransplantation trabecular bone specimen stained with toluidine blue and photographed under polarized light to enhance the woven (w) texture of the bone, a characteristic feature of patients with osteogenesis imperfecta. (F) Similarly prepared post-transplantation bone specimen demonstrating lamellar bone (L) formation, and linearly arranged osteoblasts (OB) in areas of active bone formation along the calcified trabecular surface. Magnification, 88X.

(A color version of this figure is freely accessible via the website of the book: *http://www.springer.com/1-4020-3935-2*)

Here, we will only consider the three patients in whom we documented osteoblast engraftment. Although we believe it is quite likely that the other children also had osteoblast engraftment and sampling error precluding our analysis from identifying donor cells, definitive proof of engraftment is lacking. Similarly, we do not have definitive proof of nonengraftment; therefore, those patients are inevaluable for clinical outcome.

Over the first year of life, the patients showed a typical growth pattern (Figure 8.4.A) with a decreasing growth velocity (Figure 8.4.B) similar to our controls, who were children with severe OI that did not have any specific therapy over the duration of observation. The controls exhibited the characteristic growth plateau at about 12 months of age (Figure 8.4.A), which persisted during throughout the surveillance. In contrast, all three patients showed an acute acceleration of their growth velocity during the first 6 months after BMT, which was at 13–17 months old, when the growth plateau is expected. The mean growth velocity approximated the median growth velocity for age and sex matched unaffected children. Subsequently, the growth rates slowed, but remained greater than controls. All patients showed an increase of total body bone mineral content (TBBMC) measured by dual energy X-ray absorptiometry (Figure 8.5). Most striking, one patient showed a 77% increase in TBBMC, during the first 3 months after transplant. The rate of gain in TBBMC among these patients slightly exceeded that for weight matched healthy children and the last few measurements approached the lower limit of the normal range. Although control data is not available for our measurements of TBBMC, OI is a disease of osteopenia; hence, we interpret these findings as suggestive of clinical improvement of bone mineralization. Finally, the rate of radiographically documented fractures acutely decreased during the first 6 months after transplant, compared with the 6 months before transplant. Although the rate of factures gradually declines with age, an immediate reduction in the rate of fractures is inconsistent with the natural history of OI, and most importantly, controls showed a stable rate of fractures over the age matched interval.

This trial of BMT for children with severe OI was the first prospective trial of BMT focused on nonhematopoietic cells, the first proof of engraftment of bone marrow cells in nonhematopoietic tissue, and the first proof of concept for bone marrow cell therapy of nonhematopoietic disorders. Although it represented a significant advancement in the development of cell therapy for OI and possibly other mesenchymal disorders, the children were not sufficiently improved to consider BMT, as it is currently practiced, to be a sole, complete therapeutic intervention.

4. MARROW MESENCHYMAL STEM CELL BOOSTS AS CELLULAR THERAPY FOR CHILDREN WITH SEVERE OSTEOGENESIS IMPERFECTA

4.1. Overview

In an effort to enhance the benefits observed after BMT, we developed a clinical protocol to test the hypothesis that isolated, allogeneic marrow mesenchymal stem cells could be safely infused after allogeneic BMT, and would benefit children with severe OI. To unequivocally identify the marrow mesenchymal stem cells infused in this trial and subsequent osteogenic progeny (compared to cells that may persist after the original

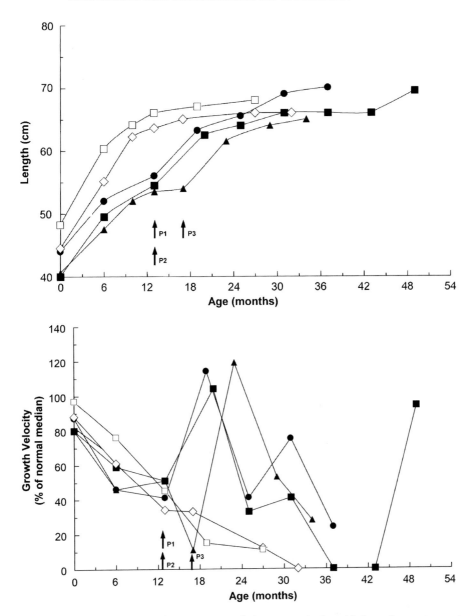

Figure 8.4. Growth profiles of OI patients and their controls from birth to the most recent assessment. (A) Absolute growth in cm. (B) Growth velocity, (difference between the first and last measurement of each interval, reported as a percentage of the median growth velocity for age- and sex- matched healthy children). Controls were children with OI who did not receive specific therapy during the observation period. Each symbol represents a crown-to-heel measurement. Filled symbols represent patients, and open symbols, controls. Arrows indicate the times of transplantation for patients (P).

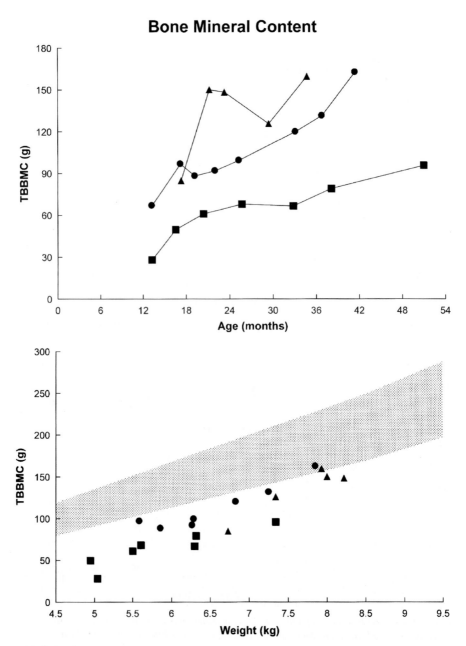

Figure 8.5. Changes in TBBMC after bone marrow transplantation. (A) Absolute measurements by dual-energy x-ray absorptiometry. (B) TBBMC as a function of body weight. The shaded area represents the normal range (±2 standard deviations from the mean) of measurements for weight-matched healthy children. Data for control patients were not available. Arrows indicate the times of transplantation. Symbols correspond to those in Figure 8.4.

BMT) we "gene marked" the cells by transduction with unique retroviral vectors. Furthermore, to investigate whether marrow mesenchymal cells could be expanded *ex vivo*, and retain their biologic potential, we used a double gene marking strategy in which minimally processed cells and expanded cells were each transduced with a distinguishable retroviral vector and infused a few weeks apart (i.e. two separate MSC infusions).

4.2. Marrow mesenchymal stem processing

A small aliquot of bone marrow was freshly harvested from the original bone marrow donors for each patient. After the MSCs were isolated from the marrow by adherence to plastic, the cells were divided into two fractions and each was transduced with one of the two retroviral vectors that may be distinguished by a PCR based assay. In this trial, one vector expressed the "marking" genetic sequence, neomycin phosphotransferase (neo^R), while the other vector did not express the encoded marking sequence. The first fraction was allowed to remain in culture for the minimal time required for isolation and transduction, while the other was expanded over three passages. The vectors used to transduce the two fractions were alternated among the patients to avoid vector bias in this marking trial. The minimally maintained cell preparation was infused into the patients, without a chemotherapy conditioning regimen, at a dose of 1×10^6 cells/kg and the expanded MSCs were infused at an intended dose of 5×10^6 cells/kg (actual median dose, 4.7×10^6) after about 2–3 weeks, again without a conditioning regimen. Contamination of the infused cells by CD45+, CD14+, or CD3+ cells generally ranged from 0 to 1.6% (median, 0.1%) without an overall difference between the minimally cultured and expanded fractions. Finally, the MSCs uniformly showed osteogenic differentiation potential when cultured in osteoinductive media. Although the MSCs were not further characterized by flow cytometry, current MSC therapy trials should include this valuable phenotyping data.

4.3. Engraftment

About 6 weeks after the cell infusions, we obtained a biopsy of bone, and skin, and an aspirate of bone marrow. Osteoblasts, skin fibroblasts, and marrow stromal cells were expanded in culture and assayed by flow cytometry to exclude lymphohematopoietic contamination. We then used our PCR assay to assess for the presence of proviral sequences in isolated DNA, indicative of engraftment of the respective MSC population. In five of the six patients enrolled in this trial, we were able to identify marked mesenchymal cells in at least one of the tissues studied. Both minimally processed cells and expanded cells engrafted, although the fraction of donor cells in any tissue, determined by PCR analysis for the vector sequences, never exceeded 1%. *Ex vivo* expansion may diminish the osteogenic engraftment and/or differentiation potential of marrow mesenchymal cells consistent with *in vitro* data [36]; however, the limited data in this trial precludes a definitive conclusion of the effect of *ex vivo* expansion.

4.4. Clinical outcome

All five children in whom we documented mesenchymal cell engraftment showed an acute acceleration of their growth velocity after MSC therapy. During the 6 months before the MSC infusions, the median growth velocity of the five patients was 20% (range, 0–40%) of that predicted for age and sex matched unaffected children. Over the first 6 months after the cell infusions, the median growth velocity was 70% (range, 60–94%). The most salient results were observed in two patients who did not measurably grow during the 6 months prior to MSC therapy, but increased their growth velocity to 67 and 94% after the infusions. There was not an unambiguous improvement of the TBBMC after the mesenchymal cell infusions. Since a chemotherapy-conditioning regimen was not given to the children prior to the cell infusions, and the cells were relatively pure compared to unmanipulated marrow (although still quite heterogeneous), the growth velocity data, TBBMC data notwithstanding, formulates a compelling argument supporting the therapeutic potential of marrow mesenchymal stem cells.

4.5. Toxicity

One patient, the child in whom marked mesenchymal cells were not identified and who did not show a clinical response, developed an urticarial rash a few minutes after completion of the second MSC infusion. The rash rapidly resolved, without sequelae, after administration of hydrocortisone and diphenhydramine. There was no clinically significant toxicity during or after the MSC infusions among the other patients.

4.6. Immunology

Marrow mesenchymal cells have been reported to be immunologically privileged [32, 37–39]. In our trial, we used two retroviral vectors, one that expressed the bacterial protein, neo^R, and one that did not express any vector encoded sequences. Interestingly, in all the patients, we found only cells marked with the nonexpressing vector. This suggested that the neo^R expressing cells were immunologically recognized when they were infused into these immunocompetent patients. In one patient, we were able to demonstrate, using a chromium release assay, cytotoxic T-cell activity against neo^R expressing mesenchymal cells in contrast to mesenchymal cells that were transduced with the nonexpressing vector or mock transduced cells. Mesenchymal cells, therefore, seem to be subject to an immune response, possibly while in the circulation, when expressing a foreign protein.

We also noted that the only child, in whom we neither identified gene marked MSC progeny nor demonstrated a clinical response, was the only child who developed an adverse event, an urticarial rash after the second MSC infusion. Urticarial rashes typically indicate a systemic immune response, which prompted us to evaluate the patients for anti-fetal bovine serum (FBS) antibodies, since FBS was a component of the media throughout the retroviral transduction and *ex vivo* expansion procedures. Using an ELISA assay, we demonstrated a greater than 100-fold increase in anti-FBS antibody titers in post-infusion serum compared to the pre-infusion serum in this patient. The remaining patients did not show a change in anti-FBS antibody titers after the infusions, which were consistent with the negative control serum in the ELISA assay. Although the lack of evidence of engraftment must be considered inconclusive as detailed above, these observations taken together, suggest that this child had anti-FBS antibodies that attacked the marrow mesenchymal cells, which precluded engraftment and thereby any clinical response. This data suggests that mesenchymal cells are subject to an immune response when presenting a foreign antigen. Most importantly, this single patient data, although unfortunate for the individual child, provides additional persuasive data supporting the therapeutic potential of MSCs.

5. CURRENT STATUS AND FUTURE PROSPECTS

Mesenchymal stem cell therapy for nonhematopoietic disorders is extraordinarily promising, but such therapy is still early in development. Our two clinical trials of bone marrow mesenchymal stem cell therapy for children with severe OI, in our view, have proven the potential of this cellular therapy for diseases of mesenchymal tissues, and bone disorders in particular. In fact, a recent paper reported that another genetic disorder of bone, hypophosphatasia, had been successfully treated with BMT and a subsequent marrow stem cell infusion [40]. Quite interestingly, the clinical course of this patient with hypophosphatasia followed a remarkably similar pattern to the OI patients in our trials. Specifically, the child did well soon after BMT, but then at 6–9 months after transplantation, the improvement slowed. Analogous to our trial schema, at 13 months post-transplantation, these investigators infused the child with a boost of *ex vivo* expanded whole bone marrow (adherent and nonadherent cells) obtained from the original marrow donor and, subsequently, the child again did well.

The comparable clinical profiles suggest marrow mesenchymal cells engraft in bone after marrow transplantation and differentiate to functional osteoblasts/osteocytes capable of improving the health of the bone; however, the engraftment is transient. At approximately 6 months after transplantation, the donor mesenchymal engraftment seems to diminish below the level of clinical significance. A second infusion of donor mesenchymal stem cells appears to engraft without further radiochemotherapy conditioning and recapitulate some of the benefits resulting from the initial marrow transplantation.

The key element to the success of mesenchymal stem cell therapy may very well lie in the development of methods that will promote long-term tissue-specific proliferation and differentiation at biologically significant levels, thus ensuring maximum clinical benefits. The ideal approach may involve the transplantation of highly purified cells, such as MSCs, as described by Pittenger [21], or a subpopulation of MSCs, such as RS cells [24], or MAPCs [26]. Conceivably, different subsets of mesenchymal cells may prove most useful for different diseases or tissues.

Ex vivo expansion of MSCs is a crucial aspect for cell processing, as cell dose may prove vital to attain sufficient levels of donor engraftment. While MSCs used in our study seem to have a finite proliferative potential, MAPCs appear to essentially have an unlimited capacity for *ex vivo* expansion [26]. Additionally, growth media is invariably supplemented with FBS, which seems to be essential for MSC proliferation *in vitro*, but can lead to an immune reaction against the processed MSCs. While serum free conditions have been described [41], technical and regulatory challenges may preclude such complex media from common use in the clinical cell processing laboratory. However, the Wnt/β-catenin signaling inhibitor Dickkopf-1 (Dkk-1), was recently reported to stimulate the undifferentiated proliferation of MSCs [42], suggesting the potential for clinically applicable serum free expansion.

We conclude that marrow mesenchymal cell therapy holds great promise for children with osteogenesis imperfecta, as well as hypophosphatasia, and likely other mesenchymal disorders. Fundamental and translational research on MSCs continues as one of the most exciting and rapidly advancing areas in biomedical science. Future discoveries of the biology of MSCs will undoubtedly foster the development of novel therapeutic interventions. Our goal is to utilize these discoveries to further refine the clinical application of bone marrow mesenchymal cell transplantation and, ultimately, to delineate a widely applicable therapy for children with genetic disorders of mesenchymal tissues.

REFERENCES

[1] Thomas ED, Lochte HL, Lu WC, Ferrebee JW. Intravenous infusion of bone marrow in patients receiving radiation and chemotherapy. *N Engl J Med*. 1957;257:491–496.

[2] Urso P, Congdon CC. The effect of the amount of isologous bone marrow injected on the recovery of hematopoietic organs, survival and body weight after lethal irradiation injury in mice. *Blood*. 1957;12:251–260.

[3] Crouch BG, Overman RR. Whole-body radiation protection in primates. *Federation Proc*. 1957;27.

[4] Ford CE, Hamerton JL, Barnes DWH, Loutit JF. Cytological identification of radiation-chimaeras. *Nature*. October 3, 1956;177:452–454.

[5] Civin CI, Strauss LC, Brovall C, et al. Antigenic analysis of hematopoiesis. III. A hematopoietic progenitor cell surface antigen defined by a monoclonal antibody raised against KG-1a cells. *J Immunol*. 1984;133:157–165.

[6] Friedenstein AJ, Petrakova KV, Kurolesova AI, Frolova GP. Heterotopic of bone marrow. Analysis of precursor cells for osteogenic and hematopoietic tissues. *Transplantation*. 1968;6:230–247.

[7] Friedenstein AJ, Gorskaja UF, Kulagina NN. Fibroblast precursors in normal and irradiated mouse hematopoietic organs. *Exp Hematol*. 1976;4:267–274.

[8] Friedenstein AJ, Latzinik NV, Gorskaya Y, Luria EA, Moskvina IL. Bone marrow stromal colony formation requires stimulation by haemopoietic cells. *Bone Miner*. 1992;18:199–213.

[9] Friedenstein AJ, Ivanov-Smolenski AA, Chajlakjan RK, et al. Origin of bone marrow stromal mechanocytes in radiochimeras and heterotopic transplants. *Exp Hematol*. 1978;6:440–444.

[10] Friedenstein AJ, Deriglasova UF, Kulagina NN, et al. Precursors for fibroblasts in different populations of hematopoietic cells as detected by the *in vitro* colony assay method. *Exp Hematol*. 1974;2:83–92.

[11] Friedenstein AJ, Chailakhyan RK, Latsinik NV, Panasyuk AF, Keiliss-Borok IV. Stromal cells responsible for transferring the microenvironment of the hemopoietic tissues. Cloning *in vitro* and retransplantation *in vivo*. *Transplantation*. 1974;17:331–340.

[12] Till JE, McCulloch EA. A direct measurement of the radiation sensitivity of normal mouse bone marrow cells. *Radiat Res*. 1961;14:213–222.

[13] Owen M. Lineage of osteogenic cells and their relationship to the stromal system. In: Peck W, ed. *J Bone Min Res*. Vol 3. Elsevier; 1985:1–25.

[14] Owen M, Friedenstein AJ. Stromal stem cells: marrow-derived osteogenic precursors. *Ciba Found Symp*. 1988;136:42–60.

[15] Caplan AI. Mesenchymal stem cells. *J Orthop Res*. 1991;9:641–650.

[16] Pereira RF, Halford KW, O'Hara MD, et al. Cultured adherent cells from marrow can serve as long-lasting precursor cells for bone, cartilage, and lung in irradiated mice. *Proc Natl Acad Sci USA*. 1995;92:4857–4861.

[17] Byers PH. Disorders of collagen biosynthesis and structure. In: Scriver CR, Beaudet AL, Sly WS, Valle D, eds. *The Metabolic and Molecular Bases of Inherited Disease*. New York, NY: McGraw-Hill; 1995:4029–4077.

[18] Sillence DO. Disorders of bone density, volume, and mineralization. In: Rimoin DL, Connor JM, Pyeritz RE, eds. *Emery and Rimoin's Principles and Practice of Medical Genetics*. 3rd ed., Vol 2. New York, NY: Churchill Livingstone; 1997.

[19] Lazarus HM, Haynesworth SE, Gerson SL, Rosenthal NS, Caplan AI. *Ex vivo* expansion and subsequent infusion of human bone marrow-derived stromal progenitor cells (mesenchymal progenitor cells): implications for therapeutic use. *Bone Marrow Transplant*. 1995;16:557–564.

[20] Horwitz EM, Keating A. Nonhematopoietic mesenchymal stem cells: what are they? *Cytotherapy*. 2000;2:387–388.

[21] Pittenger MF, Mackay AM, Beck SC, et al. Multilineage potential of adult human mesenchymal stem cells. *Science*. 1999;284:143–147.

[22] Long MW, Robinson JA, Ashcraft EA, Mann KG. Regulation of human bone marrow-derived osteoprogenitor cells by osteogenic growth factors [published erratum appears in *J Clin Invest*. 1995;96(5):2541]. *J Clin Invest*. 1995;95:881–887.

[23] Long MW, Williams JL, Mann KG. Expression of human bone-related proteins in the hematopoietic microenvironment. *J Clin Invest*. 1990;86:1387–1395.

[24] Colter DC, Class R, Digirolamo CM, Prockop DJ. Rapid expansion of recycling stem cells in cultures of plastic-adherent cells from human bone marrow. *Proc Natl Acad Sci USA*. 2000;97:3213–3218.

[25] Colter DC, Sekiya I, Prockop DJ. Identification of a subpopulation of rapidly self-renewing and multipotential adult stem cells in colonies of human marrow stromal cells. *Proc Natl Acad Sci USA*. 2001;98:7841–7845.

[26] Jiang Y, Jahagirdar BN, Reinhardt RL, et al. Pluripotency of mesenchymal stem cells derived from adult marrow. *Nature*. 2002;418:41–49.

[27] Olmsted-Davis EA, Gugala Z, Camargo F, et al. Primitive adult hematopoietic stem cells can function as osteoblast precursors. *Proc Natl Acad Sci USA*. 2003;100:15877–15882.

[28] Constantinou CO, Pack M, Young S, Prockop DJ. Phenotypic heterogeneity in osteogeneisi imperfecta: the mildly affected mother of a proband with a lethal variant has the same mutation substituting cysteine for a1-glycine 904 in a type I procollagen gene (COL1A1). *Am J Hum Genet*. 1990;47:670–679.

[29] Sokolov BP, Mays P, Khillan J, Prockop DJ. Tissue- and development specific expression in transgenic mice of a type I procollagen (COL1A1) minigene construct with 2.3 kb of the promoter region of 2 kb of the 3' flanking region. Specificty is independent of the putative regulatory sequences in the first intron. *Biochemistry*. 1993;32:9242–9249.

[30] Pereira RF, O'Hara MD, Laptev A, et al. Marrow stromal cells as a source of progenitor cells for nonhematopoietic tissues in transgenic mice with a phenotype of osteogenesis imperfecta. *Proc Natl Acad Sci USA*. 1998;95:1142–1147.

[31] Horwitz EM, Prockop DJ, Fitzpatrick LA, et al. Transplantability and therapeutic effects of bone marrow-derived mesenchymal cells in children with osteogenesis imperfecta. *Nat Med*. 1999;5:309–313.

[32] Horwitz EM, Gordon PL, Koo WKK, et al. Isolated allogeneic bone marrow-derived mesenchymal cells engraft and stimulate growth in children with osteogenesis imperfecta: implications for cell therapy of bone. Proc Natl Acad Sci USA. 2000;99(13):8932–8937.

[33] Brenner MK, Rill DR, Moen RC, et al. Gene-marking to trace origin of relapse after autologous bone-marrow transplantation. *Lancet*. 1993;341:85–86.

[34] Robey PG, Termine JD. Human bone cells *in vitro*. *Calcif Tissue Int*. 1985;37:453–460.

[35] van der PA, Aarden EM, Feijen JH, et al. Characteristics and properties of osteocytes in culture. *J Bone Miner Res*. 1994;9:1697–1704.

[36] Banfi A, Muraglia A, Dozin B, et al. Proliferation kinetics and differentiation potential of *ex vivo* expanded human bone marrow stromal cells: Implications for their use in cell therapy. *Exp Hematol*. 2000;28:707–715.

[37] Bartholomew A, Sturgeon C, Siatskas M, et al. Mesenchymal stem cells suppress lymphocyte proliferation *in vitro* and prolong skin graft survival *in vivo*. *Exp Hematol*. 2002;30:42–48.

[38] Kleeberger W, Versmold A, Rothamel T, et al. Increased chimerism of bronchial and alveolar epithelium in human lung allografts undergoing chronic injury. *Am J Pathol*. 2003;162:1487–1494.

[39] Le Blanc K, Tammik L, Sundberg B, Haynesworth SE, Ringden O. Mesenchymal stem cells inhibit and stimulate mixed lymphocyte cultures and mitogenic responses independently of the major histocompatibility complex. *Scand J Immunol*. 2003;57:11–20.

[40] Whyte MP, Kurtzberg J, McAlister WH, et al. Marrow cell transplantation for infantile hypophosphatasia. *J Bone Miner Res.* 2003;18:624–636.

[41] Gronthos S, Simmons PJ. The growth factor requirements of STRO-1-positive human bone marrow stromal precursors under serum-deprived conditions *in vitro. Blood.* 1995;85:929–940.

[42] Gregory CA, Singh H, Perry AS, Prockop DJ. The Wnt signaling inhibitor dickkopf-1 is required for reentry into the cell cycle of human adult stem cells from bone marrow. *J Biol Chem.* 2003;278:28067–28078.

CHAPTER 9

CLINICAL TRIALS OF HUMAN MESENCHYMAL STEM CELLS TO SUPPORT HEMATOPOIETIC STEM CELL TRANSPLANTATION

O. N. KOÇ

Division of Hematology-Oncology, Department of Medicine, Ireland Cancer Center, Cleveland, OH

1. INTRODUCTION

Recognition of the hematopoietic support function of the undifferentiated mesenchymal stem cells prompted clinicians to use these cells as adjuvant cellular therapy during hematopoietic stem cell transplantation. The hypothesis was to improve both the frequency and the speed of hematopoietic engraftment after hematopoietic stem cell transplantation by co transplantation of donor mesenchymal stem cells, particularly in high risk patients such as those receiving marginal numbers of hematopoietic stem cells (i.e. umbilical cord blood) and those receiving unrelated donor or related, but non-human leukocyte antigen (HLA)-identical donor stem cells. This hypothesis led to the earliest clinical trials with infusion of first, autologous and later, allogeneic human MSCs. This chapter will review the rationale of using MSCs as adjunct to the hematopoietic stem cell transplantation, results of early clinical trials and the immune properties of MSCs with a renewed interest in their clinical use to affect graft rejection and graft-versus-host disease (GVHD), two major complications of allogeneic hematopoietic stem cell transplantation.

Although the term MSC is used in this book to describe a unique cell population obtained from the bone marrow, the reader should be aware of the variability in isolation techniques of MSCs and the characteristics of MSCs used in different laboratories and clinical scale manufacturing sites. The most widely used technique adopted by the clinical trials was developed by the investigators at Case Western Reserve University (CWRU) based on the rapid plastic adherence and high proliferation potential of MSCs in 10% fetal calf serum [1, 2]. Generated cultures of a relatively uniform population of adherent cells could be used in the clinic either in undifferentiated state or after differentiation along the osteogenic, chondrogenic and adipogenic lineages.

J.A. Nolta (ed.), Genetic Engineering of Mesenchymal Stem Cells, 151–162.

2. HEMATOPOIETIC PROPERTIES OF MSC

A number of chemotherapeutic agents have been shown to be toxic to the marrow microenvironment. The ability to form a confluent stromal layer was significantly diminished in marrow specimens obtained from experimental animals and patients undergoing myeloablative radiotherapy and chemotherapy. Agents such as busulfan, cyclophosphamide and BCNU not only cause stromal damage but also diminish the ability of stroma to support hematopoiesis [3–7]. Therefore repletion of functionally competent elements of the bone marrow microenvironment is expected to improve hematopoiesis following toxic injury to the bone marrow. It is thought that mesenchymal stem cells give rise to adventitial and other mesenchymal cells in the marrow and constitute the microenvironment for hematopoiesis. Such cells fabricate the connective tissue scaffolding and produce cytokines, chemokines and extracellular matrix proteins that regulate hematopoietic homing and proliferation [8, 9].

In unstimulated cultures, MSCs appear as fusiform fibroblasts and express a number of hematopoietic cytokines, including interleukin-6 (IL-6), -7, -8, -11, -12, -14, -15, monocyte-colony stimulating factor (M-CSF), flt-3 ligand (FL), leukemia inhibitory factor (LIF) and stem cell factor (SCF) [10–12] but not IL-3. Exposure to dexamethasone results in decreased expression of LIF, IL-6 and IL-11. In contrast, IL-1α increases the expression of G-CSF, M-CSF, LIF, IL-1, IL-6, IL-8 and IL-11 and induces expression of GM-CSF but does not alter the expression of IL-7, IL-12, IL-14, IL-15, M-CSF, FL and SCF. Similar to Dexter type stromal cultures containing a more complex mixture of cells, MSCs can support human long-term culture-initiating cells (LTC-ICs) [11, 12] and *ex vivo* expansion of umbilical cord blood (UCB) hematopoietic progenitors [13].

3. IMMUNE PROPERTIES OF MSC

MSCs express a number of molecules on their surface suitable for interaction with T lymphocytes. These include VCAM-1 interacting with VLA-4, ICAM-1 interacting with LFA-1, LFA-3 interacting with CD2, and HLA MHC Class I interacting with CD8 molecules found on T-cells. Only after IFNγ treatment MHC Class II molecules were detected on MSCs and the expression of Class I molecules was enhanced. B7-1 (CD80) and B7-2 (CD86) co stimulatory molecules were not detectable on MSCs by flow cytometry. As predicted by these features, human MSCs do not present antigen. On the contrary, McIntosh at al. reported MSC mediated suppression of both primary and secondary T lymphocyte proliferation in response to allogeneic stimuli [14]. Subsequently a number of laboratories reported on the inhibitory effects of MSCs on T-lymphocyte activation and proliferation [15–18]. Di Nicola et al. suggested that the mechanism of this suppression is mediated by soluble factors, including hepatocyte growth factor and transforming growth factor-β (TGFβ) [15], but other groups have not confirmed these findings. Le Blanc et al. demonstrated dose-dependent inhibition of mixed lymphocyte cultures by addition of autologous, allogeneic or third-party MSCs [16]. T-cell activation and proliferation in response to PHA, ConA, and ProteinA was also suppressed by MSCs, indicating a non-specific effect. Tse et al. also demonstrated suppression of T-cells by MSCs, which did not appear to be mediated by IL-10, TGFβ, or PG$_{E2}$ [17]. Interestingly, despite marked suppression of T-cell activation and proliferation by MSCs, anergy was not induced. Maitra et al. used a human-interferon-γ Elispot assay to determine if human MSCs activated allo-reactive T-cells in unrelated human blood [18]. Pairs of MSCs and peripheral blood mononuclear cells derived from different adult donors were investigated. Allogeneic HLA un-matched MSCs did not activated T-cells in any of the individuals tested. Same T-cells were easily activated using allogeneic mixed lymphocyte reactions and with phytohemaglutinin (PHA). Most importantly, a significant reduction of T-cell activation occurred in mixed lymphocyte reactions performed in the presence of MSCs unrelated to either lymphocyte donor. Both human and rat MSCs were immunosuppressive while human dermal fibroblasts and murine NIH-3T3 cells were not.

The immunosuppressive effect of MSCs was shown to be mediated by soluble factors using trans-well chambers instead of direct cell contact. Interestingly, conditioned supernatant of MSCs did not have a suppressive effect but rather had a stimulatory effect on lymphocytes [18]. In fact, the immunosuppressive effects of mouse and human MSCs were shown to require an activation step that involves interaction with lymphocytes [18–20]. Regulatory cells involved in this "activation" step appear to be CD8+ and not CD4+CD25+ regulatory lymphocytes [19, 20]. Conditioned supernatant obtained from a mixture of MSCs and blood lymphocytes had a profound immunosuppressive effect. The precise mechanism and the mediators of this immunosuppressive effect are under investigation and evolving rapidly. Results of these studies may potentially have a significant impact on therapeutic potential of MSCs and the transplantation field in general.

4. PRE-CLINICAL DATA

The impact of bone marrow derived stromal cell/MSC transplantation on hematopoiesis was investigated using a variety of animal models. Anklesaria et al. showed that a bone marrow stromal cell line (GB1/6) could engraft mice pre-treated with irradiation and the donor stromal cells could facilitate hematopoietic recovery from radiation [21]. Host marrow recovery was assessed following 3 Gy total body irradiation and 10 Gy unilateral hind leg radiation with or without IV infusion of 0.1–1×10^6 GB1/6 cells. GB1/6 cells were identified only in marrow sinusoids of right hind leg (high radiation exposure) 2 months post transplant and up to 80% of the stromal cells established from transplanted mice were of donor origin. Furthermore GB1/6 transplanted mice had significantly higher cell and hematopoietic colony forming unit (CFU) recovery at 1, 2 and 3 months post transplant compared to irradiated but untransplanted mice. In utero co transplantation of stromal elements and hematopoietic cells in preimmune sheep resulted in higher level of donor hematopoiesis for up to 30 months compared to sheep not receiving stromal elements [22].

More recently several investigators showed improved human hematopoiesis in NOD-SCID mice by co transplantation of human MSCs. Noort et al. used human fetal lung derived CD34+ cells to generate MSCs and co transplanted them with a limiting number of umbilical cord blood CD34+ cells. They observed a three- to fourfold increase in the level of human hematopoietic engraftment in NOD-SCID mice given fetal lung MSCs compared to those that did not receive MSCs [23]. Angelopoulou et al. co transplanted human marrow derived MSCs with mobilized blood CD34+ cells and found enhanced human myeloid and megakaryocytic engraftment in NOD-SCID mice [24]. There was also a shift from predominantly human B-lymphocyte generation in this model to myeloid progenitor production. Maitra et al. observed increased frequency and level of human hematopoietic engraftment in mice co-transplanted with human MSCs and a limiting number of human UCB cells [18]. Almedia-Porada et al. showed enhancement of human hematopoiesis in preimmune sheep with co transplantation of human stromal cell progenitors [25]. These data provide a strong preclinical rationale for co transplantation of human hematopoietic stem cells and MSCs for purposes of improving hematopoietic engraftment in patients undergoing myeloablative treatments.

An important impediment to allogeneic hematopoietic engraftment is the major histocompatibility complex (MHC) mismatch between the donor hematopoietic progenitors and the host bone marrow microenvironment. Hematopoietic engraftment of a mismatched allogeneic donor was shown to be facilitated by MHC matched bone grafting with predominant engraftment in the bone grafts [26]. Similarly, MHC matched osteoblast or CD8+, CD3+, TCR^{neg} "facilitator cell" co transplantation was shown to improve engraftment with purified allogeneic hematopoietic progenitors [27, 28]. These data suggest that stable full or mixed donor hematopoietic chimerism can be supported by co transplantation of donor bone marrow microenvironment. Mesenchymal stem cells can potentially fulfill this goal either by direct interaction with the donor immune system or by giving rise to elements of donor bone marrow microenvironment in the host.

Few *in vivo* models exist investigating the immunosuppressive properties of MSCs. Using a baboon skin graft model, Bartholomew and co-workers showed that infusion of *ex vivo*-expanded donor (baboon) MSCs at a dose of 20×10^6 MSC/kg recipient weight prolonged time to rejection of histoincompatible skin grafts [29]. Even "third-party" baboon MSCs obtained from neither recipient nor skin graft donor appeared to suppress alloreactivity *in vivo*. Potent immunosuppressive properties of mouse MSC were also reported. Co-injection of tumor cells and mouse MSCs significantly enhanced tumor growth in immunocompetent allogeneic recipients and tumor formation was more frequent compared to the controls not receiving MSC injection. Even systemic administration of MSCs promoted tumor growth in this model [19]. These data indicate a potential role for MSCs as cellular immunosuppressive therapy in the setting of autoimmune disorders, transplanted solid organ rejection and GVHD of allogeneic stem cell transplantation. There is also reason for caution since development of malignancies and loss of graft versus tumor effect may be the serious risks associated with MSCs, although these risks are common to most if not all immunosuppressive treatments.

5. MSC TRANSPLANTATION

5.1. Bone marrow transplantation versus engineered MSC transplantation

Although the bone marrow is a rich source of MSCs, conventional allogeneic bone marrow transplantation does not result in transfer of donor MSCs or MSC derived cells into the recipient [30, 31]. These results are attributed to the inability of the conditioning regimen to ablate host marrow stroma and/or the inability of stromal progenitors to engraft. In addition, the numbers of MSCs in an average bone marrow graft may be too low, estimated around 400–1000 MSCs/kg. Report of allogeneic osteoblast engraftment in children with osteogenesis imperfecta after bone marrow transplantation suggest that certain permissive conditions such as an underlying mesenchymal defect may be necessary to achieve MSC engraftment [32].

In an attempt to achieve mesenchymal engraftment, studies were initiated to investigate the transplantation of high numbers of culture expanded murine and human MSCs. Human MSCs have a high *in vitro* proliferative potential and can expand their numbers from approximately 1500–3500 MSCs per 20 ml of bone marrow aspirate at collection to $70–700 \times 10^6$ MSCs (or $1–10 \times 10^6$/kg) at the end of expansion, which is an equivalent number of MSCs found in >1000 liters of fresh bone marrow aspirate. While such *ex vivo* expansion significantly increases the number of MSCs, this process certainly alters the biology of the cells in many ways. Although most laboratories reported the maintenance of multilineage differentiation potential of expanded MSCs, homing potential of these expanded cells was found to be significantly impaired [33].

Thus far intravenous infusion of MSCs only resulted in demonstration of few donor MSCs in various tissues of recipients. A number of factors are likely to contribute to poor MSC engraftment. First, the size and surface characteristics of MSCs may not be optimal for homing to tissues in which they can proliferate. There is light microscopy and flow cytometry evidence that culture expanded adherent MSCs are relatively large ($2–3 \times$ of granulocytes). Cell size is an important issue when cells are given directly into the vasculature. Human MSCs were shown to express $\alpha 1-3$ and $\beta 1$, $\beta 3$, $\beta 4$ integrins, ICAM 1 and 2, VCAM, L-Selectin, and CD44 (hyaluronate) but not $\alpha 4$ integrin, E-Selectin, P-Selectin, ICAM-3 and Cadherin-5, important adhesion molecules in hematopoietic stem cell homing. Second, culture expanded MSCs may have a proliferative defect. Since MSCs are generally subjected to multiple cell divisions during *ex vivo* expansion, they may approach their proliferative limit and not able to expand sufficiently in recipients. Third, the bone marrow and other tissue environments may not attract circulating MSCs through homing peptides and may not provide a survival and proliferation advantage to the transplanted cells. Some studies suggest preferential homing of MSCs to injured tissues supporting this concept [34]. There is ongoing effort to understand and optimize distribution, homing and engraftment of intravenously infused MSCs.

5.2. Autologous MSC transplantation

The ability of MSCs to support hematopoiesis both *in vitro* and in animal models sparked the interest of bone marrow transplant physicians to use MSCs as supportive care for patients undergoing stem cell transplantation. A number of studies have shown that chemotherapy and radiation damage the marrow microenvironment and diminish its hematopoietic support function [3–7]. Therefore co transplantation of MSCs could provide hematopoietic cytokines, help to establish a new bone marrow

Table 9.1. Clinical trials of culture-expanded MSC transplantation.

Source of MSCs	Setting	Objectives	Number of patients	MSC dose	Reference
Autologous	Volunteer Patients	Feasibility, Safety	15	$1-50 \times 10^6$	35
Autologous	Breast Cancer AutoPBPC Tx	Safety, Recovery	32	$1-2.2 \times 10^6$/kg	1
Allogeneic	Allo BMT or PBPC Tx HLA Matched Sibling	Safety, Recovery, GVHD	43	$1-5 \times 10^6$/kg	37
Allogeneic	Storage Disorders	Safety, Enzyme Replacement	11	$2-10 \times 10^6$/kg	36
Allogeneic	Osteogenisis Imperfecta	Safety, Bone Growth	6	$1-5 \times 10^6$/kg	38
Allogeneic	T-depleted Haploidentical PBPC Tx	Engraftment GVHD	1	1.5×10^6/kg	39
Allogeneic Third party	UCB Tx	Safety, Recovery GVHD	8	$1-10 \times 10^6$/kg	40
Allogeneic Third party	MUD BMT or PBPC Tx	Safety, Recovery, GVHD	Planned		

PBPC: Peripheral Blood Progenitor Cell, Tx: Transplantation, Allo: Allogeneic, UCB: Umbilical Cord Blood, MUD: Matched unrelated donor, GVHD: Graft versus host disease.

microenvironment and support autologous and allogeneic hematopoietic engraftment and regeneration.

Feasibility and safety of clinical scale autologous and allogeneic human MSC expansion and intravenous infusion into adult and pediatric patients have been established [1, 35–37] (Table 9.1). In a pilot study, investigators from Case Western Reserve University demonstrated the safety of *ex vivo* expansion and subsequent infusion of autologous MSCs in 15 patient volunteers [35]. These individuals had hematologic malignancies that were in remission at the time of MSC collection and infusion and were not given preparative chemotherapy. Only $1-50 \times 10^6$ total autologous MSCs were intravenously infused without any toxicity. In a subsequent phase I trial, a total of $1-2.2 \times 10^6$ autologous MSCs/kg were infused into 28 breast cancer patients to augment hematopoietic engraftment after peripheral blood progenitor cell (PBPC) transplantation [1]. Bone marrow harvest and MSC culture were performed according to an investigational new drug (IND) application with the FDA. Final cellular product characterization included flow cytometry assessment of MSC purity and viability and exclusion of microbiological contamination. Twenty patients received freshly harvested cells and 8 received cryopreserved MSCs. No toxicity was detected related to intravenous MSC infusion. Clonogenic MSCs were detected in venous blood up to 1 hour after infusion of autologous MSCs in 13 out of 21 (62%) patients, while none of the patients had detectable MSCs in the blood prior to infusion. Hematopoietic engraftment was prompt in all patients with median neutrophil recovery (>500/μl) of 8 (range: 6–11) days and platelet count recovery >20,000/μl and >50,000 unsupported of 8.5 days (range: 4–19) and 13.5 days (range: 7–44) respectively. Based on these results a randomized multi-center trial was initiated for patients undergoing PBPC transplantation for breast cancer. This trial did not achieve the accrual goal and was prematurely terminated. Ultimate the utility of MSCs in the setting of PBPC transplantation is expected to be for patients who are in need of autologous hematopoietic stem cell transplantation but have marginal numbers of stem cells to carry out the procedure safely. A number of patients with hematologic malignancies fail to mobilize sufficient CD34+ cells and this otherwise-acceptable patient group may be excluded from autotransplants. While some investigators use supplemental bone marrow harvest in addition to the limited blood stem cells, transplant-related mortality often remains high in these patients. It remains to be seen whether these "poor mobilizing" patients also will have poor MSC yields due to prior chemotherapy and stromal injury. The immunologic inertness of MSCs may afford the use of unrelated normal donor allogeneic MSCs in the context of autologous PBPC transplantation.

5.3. Allogeneic MSC transplantation

Engraftment failure remains an important risk for many patients undergoing allogeneic hematopoietic transplantation, particularly in those receiving alternative grafts such as umbilical cord blood and those obtained from haplo-identical donors. Decreasing stem cell dose, increased HLA disparity between donor and recipients and the lower intensity immunosuppression used, all contribute to the risk of engraftment failure. Donor derived MSCs may be potentially useful to solve this problem. Furthermore, the recent data on the immunosuppressive potential of MSCs suggest an added role for MSCs in overcoming HLA barriers and perhaps reducing the incidence and severity of GVHD when given during or after allogeneic stem cell transplantation.

The pilot trial of allogeneic donor MSCs was performed in patients with lysosomal storage disorders who had successfully engrafted after an HLA-identical sibling bone marrow transplantation [36]. Allogeneic bone marrow transplantation has been shown to ameliorate clinical manifestations of selected lysosomal and peroxisomal diseases by providing normal hematopoietic stem cells that can differentiate into tissue macrophages. Despite the transfer of such cells, some patients have an incomplete correction of their disorder. MSCs have been shown to express high amounts of αL-iduronidase (deficient in Hurler disease) and arylsulfatase-A (deficient in metachromatic leukodystrophy-MLD) and could have a therapeutic effect in these storage disorders [31]. In order to demonstrate feasibility and safety of allogeneic donor MSC infusions and to provide normal enzyme into tissues of 11 patients with Hurler or MLD, $2–10 \times 10^6$ normal allogeneic MSCs were intravenously infused without and preparative chemotherapy. Toxicity was limited to grade 1 fever in three patients. There was a preliminary suggestion of clinical benefit in few patients and some of the patients went on to receive repeat infusions of MSCs [41]. In this trial, donor MSCs failed to activate recipient T-cells obtained before and after MSC infusion. Horwitz et al. infused $1–5 \times 10^6$ gene-marked allogeneic donor MSCs into six patients with osteogenesis imperfecta after conventional bone marrow transplantation [38]. Gene marked cells were detected in recipient tissues 4–6 weeks later and clinical improvement was noted in five children. Interestingly, only the MSCs transduced with a marker gene that is not expressed was detected in recipients. MSCs expressing the neomycin resistance gene induced a lytic T-cell response and these cells could not be detected in tissue biopsies. This observation suggests that immunosuppressive properties of MSCs were not sufficient to prevent development of a strong T-cell response against a bacterial protein even in the setting of allogeneic transplantation. In another trial, allogeneic bone fragments were placed intraperitoneally after bone marrow transplantation followed by infusion of donor osteoblast-like cells 2 weeks later. In three out of five patients donor stromal cells could be detected in the bone marrow with correction of hypophosphatasia in one patient and resolution of an autoimmune disorder in another patient [42]. These data established the feasibility and safety of allogeneic MSC transplantation and set the stage for co transplantation of allogeneic MSCs and hematopoietic stem cells following myeloablative conditioning.

A multi-center clinical trial of hematopoietic and mesenchymal stem cell co transplantation was conducted in patients with hematological malignancies who had an HLA-identical sibling donor. The objectives of this study were to determine the rate and rapidity of hematopoietic engraftment and the incidence and severity of GVHD. Forty-three patients were infused with $1–5 \times 10^6$ allogeneic MSCs/kg 4 hours prior to the infusion of the hematopoietic graft. Due to the time restrains for donor MSC expansion, the majority of patients received 1 or 2.5×10^6 MSCs/kg. No toxicity or delay in hematopoietic recovery occurred due to MSC infusion. Engraftment rate and rapidity and the incidence of acute and chronic GVHD were reported in a preliminary fashion [37]. When compared to an historic group of age- and disease-matched patients, faster neutrophil and platelet recovery, lower rate of acute GVHD and lower rate of death were observed in patients receiving MSCs [43]. The impact of MSC co transplantation on the incidence and severity of GVHD and the graft vs. leukemia/tumor effect has yet to be tested in a randomized trial.

In addition, the effect of MSCs on hematopoietic engraftment should be tested in patients at high risk for engraftment failure such as those receiving UCB or T-lymphocyte depleted donor cells and those undergoing a non-myeloablative allogeneic transplant. Notably, successful hematopoietic engraftment was reported in a leukemia patient after T-cell depleted HLA-mismatched parental blood stem cell and donor MSC co-infusion [39]. The lack of any GVHD in this case is intriguing, however formal assessment of the effect of MSCs on GVHD and related mortality should be investigated in patients undergoing unrelated or related but non HLA-identical donor transplantation.

UCB represent a potentially attractive alternative source of hematopoietic stem cells for patients who require allogeneic stem cell transplantation. UCB is advantageous compared to other alternative donor sources since the graft is rapidly available, and the potential for GVHD in recipients may be reduced even in the setting of HLA disparity. In adults however this approach has been hampered by the small numbers of hematopoietic stem cells available in a single UCB unit. In particular, the time to neutrophil engraftment has been relatively long [44]. A phase I trial has been initiated at the University of Minnesota in which third-party MSCs are isolated from the parent or sibling and expanded to be infused into the patient at the time of unrelated UCB transplantation. Thus far eight patients received $1-10 \times 10^6$ MSCs/kg along with a single unit of UCB. Preliminary results indicate improved hematopoietic recovery compared to historic controls [40].

A major obstacle in clinical trials with allogeneic MSCs has been the length of the required culture time to obtain high numbers of cells for therapeutic purposes. There has been a frequent discrepancy in the numbers of MSCs targeted for infusion and the actual numbers infused. This discrepancy stems from the biologic differences in donors, amount and technique of the bone marrow aspirate harvest obtained as a starting material, culture methods used and the available time for culture. Many patients with hematopoietic malignances require urgent therapy and don't have the necessary time for culture of MSCs. Therefore it would be ideal if there were no need for HLA matching between the donor MSCs and the recipient, which would allow any healthy donor to provide large number of cryopreserved MSCs available for use, off the shelf. These "universal donors" would undergo vigorous screening to prevent transmission of infectious diseases and donate bone marrow aspirate repeatedly for generation of large numbers of MSCs. Based on the preclinical data generated by the Osiris Therapeutics Inc. (personal communication, K. Atkinson) indicating survival of MHC unmatched MSCs in goat, dog, pig and baboon models, several new clinical trials are about to be launched using unrelated, HLA-unmatched, "universal donor" MSCs. It remains to be seen if universal MSC infusion will be safe and the cells will survive and provide clinical benefit in the host.

6. CONCLUSION

MSCs can be generated as homogenous population of cells that can be quantified, qualitatively analyzed and *ex vivo* manipulated. MSCs have a number of biological properties that include hematopoietic support and immunosuppression and they represent excellent cellular vehicles to express and deliver therapeutic genes. During the *ex vivo* culture period, MSCs can be transduced efficiently with retroviral vectors to express genes of interest.

Clinical applications of human MSCs are evolving rapidly with ambitious goals of improving hematopoietic engraftment rate and pace, ameliorating or preventing GVHD, correcting inborn metabolic errors and delivering a variety of therapeutic genes. A number of challenging fundamental questions regarding biology and therapeutic potential of MSCs remain unanswered. The search continues for a sub-population of MSCs with *in vivo* self-renewal and engraftment potential. Size and surface characteristics of these cells are re-examined to understand barriers to their homing *in vivo*. In the clinical setting, it remains to be seen whether transplantation of MSCs has significant value. MSCs clearly have unique immunologic features and they may offer a novel paradigm of cellular therapy for immune disorders and hematopoietic and solid organ transplantation. A firm understanding of the interaction between the MSC and the immune system is paramount for successful translation of recent findings into clinical therapeutic strategies. In addition to basic research, pre-clinical animal models are needed to ask important questions such as optimal delivery, dose, timing and source of MSCs. These and other questions are likely to keep both basic scientists and clinical researchers busy during the next few decades.

REFERENCES

[1] Koç ON, Gerson S, Cooper B, et al. Rapid hematopoietic recovery after co-infusion of autologous culture-expanded human mesenchymal stem cells (hMSCs) and PBPCs in breast cancer patients receiving high dose chemotherapy. *J Clin Oncol.* 2000:18:307–316.

[2] Lennon DP, Haynesworth SE, Bruder SP, Jaiswal N, Caplan AI. Development of a serum screen for mesenchymal progenitor cells from bone marrow. *In Vitro Cell Dev Biol.* 1996;32:602–611.

[3] Galotto M, Berisso G, Delfino L, et al. Stromal damage as consequence of high-dose chemo/radiotherapy in bone marrow transplant recipients. *Exp Hematol.* 1999;27:1460–1466.

[4] Barbot C, Rice A, Vanes I, Mahon FX, Jazwiec B, Reiffers J. Quality and functional capacity of the bone marrow microenvironment of autologous blood stem cell transplantation recipients. *Nouv Rev Fr Hematol.* 1994;36:325–331.

[5] Fried W, Kedo A, Barone J. Effects of cyclophosaphamide and of busulfan on spleen colony-forming units and on hematopoietic stroma. *Cancer Res.* 1977;37:1205–1209.

[6] McManus PM, Weiss L. Busulfan-induced chronic bone marrow failure: changes in cortical bone, marrow stromal cells, and adherent cell colonies. *Blood.* 1984;64:1036–1041.

[7] Uhlman DL, Verfaillie C, Jones RB, Luikart SD. BCNU treatment of marrow stromal monolayers reversibly alters haematopoiesis. *Br J Haematol.* 1991;78:3304–3309.

[8] Toksoz D, Zsebo K, Smith K, et al. Support of human hematopoiesis in long term bone marrow cultures by murine stromal cells selectively expressing the membrane bound and secreted forms of the human homologue of the steel gene product, stem cell factor. *Proc Natl Acad Sci USA.* 1992;89:7350–7354.

[9] Clark B, Keating A. Biology of bone marrow stroma. *Ann NY Acad Sci.* 1995;770:70–88.

[10] Haynesworth S, Baber M, Caplan A. Cytokine expression by human marrow derived mesenchymal progenitor cells *in vitro*: effects of dexamethasone and IL-1a. *J Cell Physiol.* 1996;166:585–592.

[11] Majumdar MK, Thiede MA, Mosca JD, Moorman M, Gerson SL. Phenotypic and functional comparison of cultures of marrow-derived mesenchymal stem cells (MSCs) and stromal cells. *J Cell Physiol.* 1998;176:186–192.

[12] Majumdar M, Thiede M, Haynesworth S, Bruder S, Gerson S. Human marrow-derived mesenchymal stem cells (MSCs) express hematopoietic cytokines and support long-term hematopoiesis when differentiated toward stromal and osteogenic lineages. *J Hematother Stem Cell Res.* 2001;9:841–848.

[13] Kadereit S, Deeds L, Haynesworth S, et al. Mesencymal stem cell as a feeder layer during short-term umbilical cord blood *in vivo* expansion increases early hematopoietic progenitors. *Stem Cells.* 2002;20:573–582.

[14] Klyushnenkova E, Mosca J, McIntosh K. Human mesenchymal stem cells suppress allogeneic T cell responses *in vitro*: implications for allogeneic transplantation. *Blood.* 1998;92:642a.

[15] Di Nicola M, Carlo-Stella C, Magni M, et al. Human bone marrow stromal cells suppress T-lymphocyte proliferation induced by cellular or nonspecific mitogenic stimuli. *Blood.* 2002;99:3838–3843.

[16] LeBlanc K, Tammik L, Sundberg B, Haynesworth S, Ringden O. Mesenchymal stem cells inhibit and stimulate mixed lymphocyte cultures and mitogenic responses independently of the major histocompatibility complex. *Scan J Immunol.* 2003;57:11–20.

[17] Tse WT, Pendleton JD, Beyer WM, Egalka MC, Guinan EC. Suppression of allogeneic T-cell proliferation by human marrow stromal cells: implications in transplantation. *Transplantation.* 2003;75:389–397.

[18] Maitra B, Szekely E, Gjini K, Laughlin MJD, Haynesworth S, Koç ON. Human mesenchymal stem cells support unrelated donor hematopoietic stem cells and suppress T-cell activation. *Bone Marrow Transplant.* 2003.

[19] Farida D, Pascale P, Claire B, et al. Immunosuppressive effect of mesenchymal stem cells favors tumor growth in allogeneic animals. *Blood.* 2003.

[20] Krampera M, Glennie S, Dyson J, et al. Bone marrow mesenchymal stem cells inhibit the response of naive and memory antigen-specific T cells to their cognate peptide. *Blood.* 2003;101:3722–3729.

[21] Anklesaria P, Kase K, Glowacki J, et al. Engraftment of a clonal bone marrow stromal cell line *in vivo* stimulates hematopoietic recovery from total body irradiation. *Proc Natl Acad Sci USA.* 1987;1987:7681–7685.

[22] Almeida-Porada G, Flake A, Hudson A, Zanjani E. Co transplantation of stroma results in enhancement of engraftment and early expression of donor hematopoietic stem cells in utero. *Exp Hematol.* 1999;27:1569–1575.

[23] Noort WA, Kruisselbrink AB, Anker PS, et al. Mesenchymal stem cells promote engraftment of human umbilical cord blood-derived CD34(+) cells in NOD/SCID mice. *Exp Hematol.* 2002;30:870–878.

[24] Angelopoulou M, Novelli E, Grove J, et al. Co transplantation of human mesenchymal stem cells enhances human myelopoiesis and megakaryocytopoiesis in NOD/SCID mice. *Exp Hematol.* 2003;31:413–420.

[25] Almeida-Porada G, Porada C, Tran N, Zanjani E. Co transplantation of human stromal cell progenitors into preimmune ftal sheep results in early appearance of human donor cells in circulation and boosts cell levels in bone marrow at later time points after transplantation. *Blood.* 2000;95:3620–3627.

[26] Ishida T, Inaba M, Hisha H, et al. Requirement of donor-derived stromal cells in the bone marrow for successful allogeneic bone marrow transplantation. Complete prevention of recurrence of autoimmune diseases in MRL/MP-Ipr/Ipr mice by transplantation of bone marrow plus bones (stromal cells) from the same donor. *J Immunol.* 1994;152:3119–3127.

[27] El-Badri N, Wang BC, Good R. Osteoblasts promote engraftment of allogeneic hematopoietic stem cells. *Exp Hematol.* 1998;29:110–116.

[28] Kaufman C, Colson Y, Wren S, Watkins S, Simmons R, Ildstad S. Phenotypic characterization of a novel bone marrow-derived cell that facilitates engraftment of allogeneic bone marrow stem cells. *Blood.* 1994;84:2436–2446.

[29] Bartholomew A, Sturgeon C, Siatskas M, et al. Mesenchymal stem cells suppress lymphocyte proliferation *in vitro* and prolong skin graft survival *in vivo. Exp Hematol.* 2002;72:1653–1655.

[30] Simmons PJ, Przepiorka D, Thomas ED, Torok-Storb B. Host origin of marrow stromal cells following allogeneic bone marrow transplantation. *Nature.* 1987;328:429–432.

[31] Koç ON, Peters C, Raghavan S, et al. Bone marrow derived mesenchymal stem cells of patients with lysosomal and peroxisomal storage diseases remain host type following allogeneic bone marrow transplantation. *Exp Hematol.* 1999;27:1675–1681.

[32] Horwitz EM, Prockop DJ, Fitzpatrick LA, et al. Transplantability and therapeutic effects of bone marrow-derived mesenchymal cells in children with osteogenesis imperfecta. *Nat Med.* 1999;5:309–313.

[33] Rombouts W, Ploemacher R. Primary murine MSC show highly efficient homing to the bone marrow but lose homing ability following culture. *Leukemia.* 2003;17:160–170.

[34] Mackenzie T, Flake A. Human mesenchymal stem cells persist, demonstrate site-specific multipotential differentiation, and are present in sites of wound healing and tissue regeneration after transplantation into fetal sheep. *Blood Cells Mol Dis.* 2001;27:601–604.

[35] Lazarus H, Haynesworth S, Gerson S, Rosenthal N, Caplan A. *Ex vivo* expansion and subsequent infusion of human bone marrow-derived stromal progenitor cells (mesenchymal progenitor cells): implications for therpeutic use. *Bone Marrow Transplant.* 1995;16:557–564.

[36] Koç ON, Day J, Nieder M, Gerson S, Lazarus H, Krivit W. Allogeneic mesenchymal stem cell infusion for treatment of metachromatic leukodystrophy (MLD) and Hurler syndrome (MPS-IH). *Bone Marrow Transplant.* 2002;30:215–222.

[37] Lazarus H, Curtin P, Devine S, McCarthy P, Holland K, Moseley AAB. Role of mesenchymal stem cells in allogeneic transplantation: early phase I clinical results. *Blood.* 2000;96:392a.

[38] Horwitz E, Gordon P, Koo W, et al. Isolated allogeneic bone marrow-derived mesenchymal cells engraft and stimulate growth in children with osteogenesis imperfecta: implications for cell therapy of bone. *Proc Natl Acad Sci USA.* 2002;99:8932–8937.

[39] Lee S, Jang J, Cheong J, et al. Treatment of high-risk acute myelogenous leukaemia by myeloablative chemoradiotherapy followed by co-infusion of T cell-depleted haematopoietic stem cells and culture-expanded marrow mesenchymal stem cells from a related donor with one fully mismatched human leucocyte antigen haplotype. *Br J Haematol.* 2001;118:1128–1131.

[40] MacMillan L, Ramsay N, Atkinson K, Wagner J. *Ex vivo* culture-expanded parental haploidentical mesenchymal stem cells promote engraftment in recipients of unrelated donor umbilical cord blood: results of a phase I-II clinical trial. *Blood.* 2002;100:836a.

[41] Krivit W, Day J, Koc ON, Atkinson K. Multiple mesenchymal stem cell (allogen) infusions in patients with metachromatic leukodystrophy: safety of infusions and stability of nerve conduction velocities. *Blood.* 2002;100:524a.

[42] Cahill R, Jones O, Mueller T, El-Badri N, Good R. Replacement of recipient stromal/mesenchymal cells after bone marrow transplant using bone fragments and cultured osteoblast like cells. *Blood.* 2002;100:63a.

[43] Frassoni F, Labopin M, Bacigalupo A, et al. Expanded mesenchymal stem cells (MSC), co-infused with HLA identical hemopoietic stem cell transplants, reduce acute and chronic graft versus host disease: a matched pair analysis. *Bone Marrow Transplant.* 2002;29:S2.

[44] Laughlin MJ, Barker J, Bambach B, et al. Hematopoietic engraftment and survival in adult recipients of umbilical- cord blood from unrelated donors. *N Engl J Med.* 2001;344:1815–1822.

INDEX